Fossil Ecosystems of North America: A Guide to the Sites and their Extraordinary Biotas

John R. Nudds BSc PhD
University of Manchester, Manchester, UK

Paul A. Selden BSc PhD
University of Kansas, Kansas, USA

CRC Press
Taylor & Francis Group
Boca Raton London New York

CRC Press is an imprint of the
Taylor & Francis Group, an **informa** business

Dedication

To Chris and Maura

First published 2008 by Manson Publishing Ltd

Published 2018 by CRC Press
Taylor & Francis Group
6000 Broken Sound Parkway NW, Suite 300
Boca Raton, FL 33487-2742

First issued in hardback 2018

ISBN 13: 978-1-138-42408-1 (hbk)
ISBN 13: 978-1-84076-088-0 (pbk)

Visit the Taylor & Francis Web site at
http://www.taylorandfrancis.com

and the CRC Press Web site at
http://www.crcpress.com

Cover and book design: Cathy Martin, Presspack Computing Ltd
A CIP catalogue record for this book is available from the British Library.

Contents

Acknowledgements

Photography and Illustrations

American Museum of Natural History Library: 147, 150.

Stan Awramik, University of California, Santa Barbara, USA: 11.

Black Hills Institute for Geological Research, Hill City, South Dakota, USA: 167, 170, 173, 175.

Fred Broadhurst, University of Manchester, UK: 108, 109, 111, 112, 113, 114, 115, 116, 117, 118, 119, 120, 121, 122, 123, 124, 125, 126, 127.

Sam Ciurca, Rochester, New York, USA: 72, 75, 81, 85.

Simon Conway Morris, University of Cambridge, UK: 34, 36, 37, 45, 46.

Pat Craig, Monte Rio, California, USA: 252, 253, 254, 255, 256, 257, 258, 259, 260, 261, 262, 263, 264, 265, 266, 267, 268, 269, 270, 271, 272, 273, 274, 275, 276, 277, 278, 282, 283.

Dianne Edwards, University of Wales, Cardiff, UK: 84.

Una Farrell, Yale Peabody Museum, New Haven, Connecticut, USA: 53, 60, 61, 62.

Jim Gehling, South Australian Museum, Adelaide, Australia: 27.

David Green, University of Manchester, UK: 7, 39.

Richard Hartley, University of Manchester, UK: 1, 4, 15, 18, 28, 31, 106, 107, 110, 128, 129, 133, 142, 144, 163, 166, 187, 189, 284, 287.

Linda Hernick, New York State Museum, Albany, New York, USA: 89.

Ken Higgs, University College Cork, Ireland: 288.

Cindy Howells, National Museum of Wales, Cardiff, UK: 141, 191, 199, 204.

Barry James, Prehistoric Journeys, Sunbury, Pennsylvania, USA: 177.

Andrew Knoll, Harvard University, Cambridge, Massachusetts, USA: 8, 9, 10, 13, 14.

Herbert Meyer, National Park Service, Colorado, USA: 218.

Natural History Museum of Los Angeles County, USA: 289, 292, 294, 301.

John Nudds, University of Manchester, UK: 2, 3, 5, 6, 16, 17, 19, 20, 21, 24, 25, 29, 30, 130, 131, 132, 139, 140, 143, 145, 146, 154, 157, 158, 160, 164, 165, 168, 179, 181, 182, 183, 184, 186, 188, 190, 192, 193, 194, 195, 196,197, 198, 200, 201, 202, 203, 205, 206, 207, 210, 213, 214, 285, 286, 297, 299.

David Penney, University of Manchester, UK: 250, 279, 280, 281.

Burkhard Pohl, Wyoming Dinosaur Center, USA: 156.

Graham Rosewarne, Avening, Gloucestershire, UK: 12, 22, 26, 32, 33, 35, 38, 40, 41, 42, 44, 47, 135, 136, 137, 138, 149, 152, 155, 159, 162, 169, 171, 172, 174, 176, 178, 180, 185, 209, 290, 291, 293, 295, 296, 298, 300.

Sauriermuseum, Aathal, Switzerland: 148, 151, 161.
Paul Selden, University of Kansas, USA: Table 1, 48, 50, 51, 54, 63, 64, 65, 66, 67, 68, 69, 70, 71, 73, 74, 76, 77, 78, 79, 80, 82, 83, 86, 87, 88, 90, 91, 92, 93, 94, 99, 101, 102, 134, 139, 211, 212, 215, 216, 217, 219, 220, 221, 222, 223, 224, 225, 226, 227, 228, 229, 230, 231, 232, 233, 234, 235, 236, 237, 238, 239, 240, 241, 242, 243, 244, 245, 246, 247, 248, 249, 251.
Tom Sharpe, National Museum of Wales, Cardiff, UK: 23.
Bill Shear, Hampden-Sydney College, Virginia, USA: 95, 96, 97, 98, 100, 103, 104, 105.
Jim Tynsky, Tynsky's Fossils, Wyoming, USA: 208.
The University of Wyoming Geological Museum, USA: 153.
Tom Whiteley, Rochester, New York, USA: 49, 52, 55, 56, 57, 58, 59.
Harry Whittington, University of Cambridge, UK: 43.

Access to sites and other help

Lindsey Axe, Doug Boyce, Brent Breithaupt, Derek and Jenny Briggs, Dan Brinkman, Susan Butts, Sam Ciurca, Pat Craig, Dallas Evans, Bob Farrar, Una Farrell, Mike Flynn, Richard Fortey, Rick Hebdon, Linda Hernick, Cindy Howells, Walter Joyce, Paul Kenrick, Stan Koziarz, Neal Larson, Cape MacClintock, Cathy McNassor, Herb Meyer, Urs Möckli, Dave Palling, Dave Penney, Burkhard Pohl, Victor Porter, Glenn Rockers, Andrew Ross, Pat Rusz, Russell Shapiro, Tom Sharpe, Chris Shaw, Bill Shear, Erik Tetlie, Rene Vandervelde, Katherine Ward, Tom Whiteley.

Preface

Most major advances in understanding the history of life on Earth in recent years have been through the study of exceptionally well preserved biotas (Fossil-Lagerstätten). Indeed, particular Fossil-Lagerstätten, such as the Cambrian Burgess Shale of British Columbia and the Jurassic dinosaurs of the Morrison Formation, have gained exceptional fame through popular science writings. Study of a selection of such sites scattered throughout the geological record – windows on the history of life on Earth – can provide a fairly complete picture of the evolution of ecosystems through time.

This book follows from an earlier one which covered Fossil-Lagerstätten around the world (Selden and Nudds, 2004). The first book arose from the realization that there was an obvious void in the range of paleontology texts available for a book which brings together succinct summaries of most of the better known Fossil-Lagerstätten, primarily for a student and interested amateur readership. The authors taught an undergraduate course at third-year level which was based on case studies of a number of Fossil-Lagerstätten, and also collaborated on the design of a new fossil gallery at Manchester University Museum based around this theme. The success of the first text prompted the present work, which draws four localities from the original book and adds another ten; all are located in North America (broadly – we include the USA, Canada, and the Dominican Republic).

Following an introduction to Fossil-Lagerstätten and their distribution through geological time, each chapter deals with a single fossil locality. Each chapter follows the same format: after a brief introduction placing the Lagerstätte in an evolutionary context, there then follows a history of study of the locality, the background sedimentology, stratigraphy and paleoenvironment; a description of the biota; discussion of the paleoecology; and a comparison with other Lagerstätten of a similar age and/or environment. At the end of the book is an Appendix listing museums in which to see exhibitions of fossils from each locality and suggestions for visiting the sites. We encourage the reader to visit museums and see the fossils first-hand, and to go to the localities, where this is possible, to see the fossils in the rock; in some cases you can collect your own!

Abbreviations

Specimen repositories are abbreviated in the figure captions as follows:

AMNH	American Museum of Natural History, New York, USA
AMPP	Museo del Ámbar Dominicano, Puerto Plata, Dominican Republic
BHIGR	Black Hills Institute of Geological Research, Hill City, South Dakota, USA
CFM	Field Museum of Natural History, Chicago, USA
FBNM	Fossil Butte National Monument Visitor Center, Kemmerer, Wyoming, USA
GCPM	George C. Page Museum, Los Angeles, USA
GMC	Geological Museum, Copenhagen, Denmark
GSC	Geological Survey of Canada, Ottawa, Canada
MCZ	Museum of Comparative Zoology, Harvard University, USA
MM	Manchester University Museum, UK
MU	Manchester University, UK
NHM	Natural History Museum, London, UK
PC	Private Collection
SMA	Sauriermuseum Aathal, Switzerland
TFSK	Tynsky's Fossil Shop, Kemmerer, Wyoming, USA
USNM	US National Museum of Natural History, Smithsonian Institute, Washington DC, USA
UWGM	University of Wyoming Geological Museum, Laramie, Wyoming, USA
WDC	Wyoming Dinosaur Center, Thermopolis, Wyoming, USA
YPM	Yale Peabody Museum, New Haven, Connecticut, USA

Introduction

The fossil record is very incomplete. Only a tiny percentage of plants and animals alive at any one time in the past get preserved as fossils, so that the paleontologist attempting to reconstruct ancient ecosystems is, in effect, trying to complete a jigsaw puzzle without the picture on the box lid and for which the majority of pieces are missing. Under normal preservational conditions, probably only around 15% of organisms are preserved. Moreover, the fossil record is biased in favour of those animals and plants with hard, mineralized shells, skeletons, or cuticle, and towards those living in marine environments. Thus, the preservational potential of a particular organism depends on two main factors: its constitution (better if it contains hard parts) and its habitat (better if it lives in an environment where sedimentary deposition occurs).

Occasionally, however, the fossil record presents us with surprises. Very rarely, exceptional circumstances of one sort or another allow unusual preservation of soft parts of organisms, or in environments where fossilization rarely happens. Rock strata within the geological record, which contain a much more completely preserved record than is normally the case, are windows on the history of life on Earth. They have been termed Fossil-Lagerstätten (Seilacher *et al.*, 1985), a name derived from the German mining tradition to denote a particularly rich seam.

There are two main types of Fossil-Lagerstätten. *Concentration Lagerstätten* (Konzentrat-Lagerstätten), as the name suggests, are simply deposits in which vast numbers of fossils are preserved, such as coquinas (shell accumulations), bone beds, cave deposits, and natural animal traps. The quality of individual preservation may not be exceptional, but the sheer numbers are informative. *Conservation Lagerstätten* (Konservat-Lagerstätten), on the other hand, preserve quality rather than quantity and this term is restricted to those rare instances where peculiar preservational conditions have allowed even the soft tissue of animals and plants to be preserved, often in incredible detail. (It may also be used for deposits that yield articulated skeletons without soft tissue.) Most of the Lagerstätten described in this book are examples of Conservation Lagerstätten.

There are various types of Conservation Lagerstätten including conservation traps such as entombment in amber, deep-freezing in permafrost, pickling in oil swamps, and mummification by desiccation. On a larger scale are *obrution deposits*, where episodic smothering ensures rapid burial of mainly benthonic (sea-floor) communities, and *stagnation deposits*, where anoxic (low oxygen) conditions in stagnant or hypersaline (high salinity) bottom waters ensures reduced microbial decay, in predominantly pelagic (open-sea) communities. In fact, most Conservation Lagerstätten combine obrution and stagnation in the preservation of soft tissue.

Taphonomy is the name given to the process of preservation of a plant or animal as a fossil. It actually consists of two main processes: *biostratinomy*, which covers the course of events from death to burial in sediment (or entombment in amber, cave deposits, and so on). The time taken by this process can vary from a few minutes (e.g. insects trapped in amber, or mammals in tar) to many years for an accumulation of bones or shells. Ideally, for exceptional preservation of soft tissues, the time between death and isolation from oxygen and decaying organisms should be short. Following burial, the process of *diagenesis* begins: the conversion of soft sediment or other deposits to rock. Further destruction of organic molecules can occur during diagenesis; the action of heat can turn organic molecules into oil and gas, for example, and crushing in coarse sand can fragment plant and animal cuticles.

Soft tissue preservation has three important implications. First, the study of soft part morphology, alongside the morphology of the shell or skeleton, allows better comparison with extant forms and provides additional phylogenetic information. Second, it enables the preservation of animals and plants which are entirely soft-bodied and which would normally stand no chance of fossilization. For example, it has been estimated that 85% of the Burgess Shale genera (Chapter 3) are entirely soft-bodied and are therefore absent from Cambrian biotas preserved under normal taphonomic conditions. The third implication follows – such Conservation Lagerstätten therefore preserve for the paleontologist a complete (or much more nearly complete) ecosystem. Comparison of such horizons in a chronological framework gives us an insight into the evolution of ecosystems over geologic time.

The Lagerstätten described in this book are arranged in chronological order, from the Archean Gunflint Chert biota to the Pleistocene Rancho La Brea (**Table 1**), so it is possible to follow the development of the Earth's ecosystems through a series of snapshots of life at a number of points in time (Bottjer *et al.*, 2002; Selden and Nudds, 2004). Whilst this does not give a complete picture of the evolving biosphere, Lagerstätten are important because they preserve far more of the biota than occurs under normal preservational conditions, so the paleontologist can see as completely as possible the ecological interactions of the organisms in that particular habitat. In the Archean, for example, we are seeing some of the very first, single-celled life forms on Earth, and in the Proterozoic Mistaken Point biota, we may be looking at a different grade of organismal organization and life mode than we see in later, Phanerozoic time. The Mistaken Point biota existed before hard parts of animals evolved and predation became widespread. By the middle Cambrian Period, almost all animal phyla had developed, and it is possible to reconstruct the ecological dynamics of the sea floor, complete with predators, scavengers, filter feeders, and deposit feeders. The Burgess Shale preserves mainly benthos (sea-floor dwellers). Benthos can be divided into infauna (animals living within the sediment) and epifauna (animals living on the sediment surface). Some of the biota in the Burgess Shale belongs to the nekton (swimmers) and some is plankton (floaters).

In contrast to the Burgess Shale, the Ordovician Beecher's Trilobite Bed contains a very selective biota preserved under rather unusual circumstances. By the time we reach the Silurian, sea life was well established in the oceans, but what makes the Bertie Waterlime special is its abundance of fossils of animals which lived in a rather restricted marine lagoon

The Geological Column showing stratigraphic positions of the Fossil-Lagerstätten described in this book

Million years before present	Era	Period			Lagerstätte
	CENOZOIC	Quaternary	Holocene Pleistocene		Rancho La Brea
1.8		Neogene	Pliocene Miocene	Tertiary	Dominican Amber
23		Paleogene	Oligocene Eocene Paleocene	Tertiary	Florissant Green River Formation
65	MESOZOIC	Cretaceous			Hell Creek Formation
145		Jurassic			Morrison Formation
200		Triassic			Chinle Group
251	PALEOZOIC	Permian			
299		Pennsylvanian (Upper Carboniferous)			Mazon Creek
318		Mississippian (Lower Carboniferous)			
359		Devonian			Gilboa
416		Silurian			Bertie Waterlime
444		Ordovician			Beecher's Trilobite Bed
488		Cambrian			Burgess Shale
542	PROTEROZOIC	PRECAMBRIAN			Mistaken Point Gunflint Chert
2500	ARCHEAN	PRECAMBRIAN			
4600					

situation or were washed in from the surrounding land; such biota are rather rare in the fossil record. A major evolutionary advance which occurred in the mid-Paleozoic was the colonization of land by plants and animals. The Devonian Gilboa biota was one of the first to be discovered in North America and is still the best-known, preserving some of the earliest land plants and animals. By Pennsylvanian times, the land in tropical regions had become well-colonized by forest, with its accompaniment of insects and their predators. The Mazon Creek biota preserves a forest ecosystem mingled with marine and nonmarine aquatic organisms in a deltaic setting, so common in this period of major coal formation. The end of the Permian Period saw the demise of some 80% of living things in the greatest mass extinction of all time.

Three Lagerstätten are represented here from the Mesozoic Era, all terrestrial. The Chinle Group of Arizona is famous for its preservation of the Petrified Forest, but in that forest lived early dinosaurs and other land animals such as insects. Dinosaurs came to dominate the scene on land during the Jurassic Period, and the Morrison Formation of the western USA is the best-known Lagerstätte preserving these giants. During the Cretaceous Period, while sea covered much of the mid-West, dinosaurs flourished on land which is now the Rockies and their foothills, including such famous beasts as *Tyrannosaurus* and *Triceratops*, found at Hell Creek. But the end of the Cretaceous Period, and the end of the Mesozoic Era, was marked by another extinction event in which the dinosaurs died out together with ammonites, marine reptiles, and some other plant and animal groups.

In the following Cenozoic Era, mammals became the dominant vertebrate group on land; bony fishes swam in rivers, lakes, and the sea; the forests, now dominated by flower-bearing trees, swarmed with modern-looking insects. The Green River Formation and the Florissant Formation, both situated in the Rocky Mountain foothills, preserve vast quantities of these fossils. Amber (fossilized tree resin) acts as a sticky trap for insects and their predators and, in a similar manner, the tar pits of Rancho La Brea attracted mammals and birds in search of a drink, trapping them and their predators and scavengers in sticky tar. Rancho La Brea preserves a snapshot of land life in southern California over the last 40,000 years.

Further Reading

Bottjer, D. J., Etter, W., Hagadorn, J. W. and Tang, C. M. (eds) 2002. *Exceptional Fossil Preservation.* Columbia University Press, New York.

Seilacher, A., Reif, W-E. and Westphal, F. 1985. Sedimentological, ecological, and temporal patterns of fossil Lagerstätten. *Philosophical Transactions of the Royal Society of London B* **311**, 5–23.

Selden, P. and Nudds, J. 2004. *Evolution of Fossil Ecosystems.* Manson, London.

The Gunflint Chert

Chapter 1

Background: The origin of life and the first 3 billion years

Life on Earth arose some 3.5 billion years ago. There is some debate concerning what actually constitutes 'life' and, indeed, whether life actually arose on this planet or originated extraterrestrially in a simple form and evolved further here. Nevertheless, the earliest fossil evidence of single-celled prokaryotes (simple cells with no nucleus or organelles) akin to modern cyanobacteria (previously known as blue-green algae) comes from 3,500 million year-old cherts in Western Australia and South Africa. For nearly 2,500 million years after its origin, life evolved very slowly and planet Earth was home to nothing but bacteria, although there was a huge array of different types, which actually constitute two separate kingdoms, the Eubacteria and the Archaebacteria. The former are mostly photosynthesizing bacteria, but many of the latter are unusual forms able to survive in extreme conditions, such as the 'hyperthermophiles' which can live in water as hot as 100°C, and the 'chemophiles' which thrive in a cocktail of toxic chemicals. Both these sets of conditions are found around submarine hydrothermal vents (or 'black smokers') and it has been suggested that life may have originated in such a setting, in which case the archaebacteria are our oldest ancestors.

Eukaryotes (cells which contain a nucleus and organelles) eventually evolved from prokaryotic bacteria about 1,200 million years ago (see p. 23) and within little more than half a billion years had diversified into both single-celled and multicellular forms constituting the four eukaryotic kingdoms known today: plants, fungi, protists, and eventually animals (Chapter 2). It is, however, with their prokaryotic bacterial ancestors that this chapter is concerned.

The fossil record so far back in deep time is sparse indeed and we rely on very occasional glimpses of these former worlds to build up some understanding of these early microbial ecosystems. To date most of the available evidence has come from two types of deposit – stromatolites and chert – quite distinct, but often closely associated. Stromatolites are laminated deposits, usually found in carbonates, and formed when successive layers of bacterial (microbial) mats were covered by sediment only to generate again before being smothered once more. Although the microbial mats have long since decayed, the laminated sediments are a clue to their former existence. Living stromatolites can still be seen on Earth today in just a handful of locations, most famously at Shark Bay in Western Australia, not far from their 3,500 million year-old cousins. Chert, on the other hand, is an extraordinarily hard rock similar to flint, which is made up of tiny interlocking crystals of quartz (silica or SiO_2). It is both resistant enough to withstand deformation and impermeable enough not to be corroded, and sometimes contains within it exquisitely preserved fossils of these early bacterial microbes.

Both stromatolites and fossiliferous cherts are scattered pretty thinly through the Precambrian, and some doubt has recently been cast on the affinities of the 3,500 million-year-old examples (see p. 23). It is, therefore, to the next oldest, universally accepted candidate, the 2,000 million-year-old Gunflint Chert of northwestern Ontario, that we turn for a glimpse of these early microbial ecosystems. It is a sobering thought, however, as Andrew Knoll (2003) points out in his thought-provoking book, *Life on a Young Planet*, that the Gunflint Chert is almost as distant in time from its Australian and South African predecessors as we are from Gunflint.

History of discovery of the Gunflint Chert

In 1953 Stanley Tyler, an economic geologist from the University of Wisconsin, was undertaking a major exploration of the Precambrian iron deposits of Minnesota and Ontario. He had traced these rocks from the Biwabik Formation of Minnesota, where the ironstone is mined in enormous open pits in the Mesabi Range, to the Gunflint Formation of Ontario, which crosses the US–Canada boundary at Gunflint Lake, extends to Kakabeka Falls near Thunder Bay and, finally, to isolated outcrops near Schreiber on Lake Superior (**1**).

The Gunflint Formation comprises alternating bands of ironstone (in the form of hematite, an iron oxide) and chert which may be red, yellow, or black. The cherts themselves often include layered stromatolites and the red variety, jasper, is particularly characteristic of the Gunflint (**7**). Stanley Tyler, however, was more interested in the jet black variety and

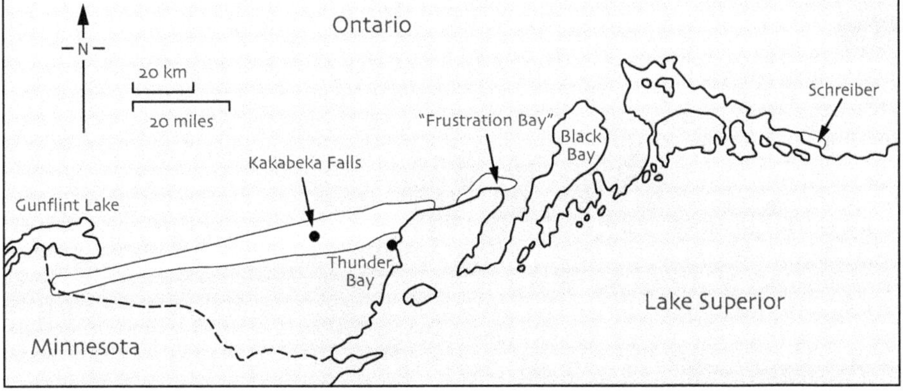

1 Locality map to show the extent of the outcrop of the Gunflint Chert (after Barghoorn and Tyler, 1965).

collected samples of these from Kakabeka Falls at Thunder Bay (**2**) and from the shores of Lake Superior near Schreiber (**3**). When he examined thin sections of this chert back in his laboratory, he was amazed to find that they contained thousands of minute spheres and stars, with branched and unbranched segmented filaments all less than 10 microns in size. He wondered if these could possibly be three-dimensionally preserved fossils, and realized that if this were the case they would be by far the oldest fossils then known from anywhere on Earth. Reg Sprigg had only discovered the 600 million year-old Ediacara fossils in 1946 (Chapter 2) and the Gunflint Chert was more than three times as old.

Tyler immediately called in paleobotanist Elso Barghoorn (1915–1984), Fisher Professor of Natural History at Harvard University, who confirmed that they were ancient fossils of unicellular organisms. The two scientists published their discovery in *Science* in 1954 (Tyler and Barghoorn, 1954) describing the fossils as 'algae and fungi' and comparing their preservation to plants from the well-known Rhynie Chert

2 Kakabeka Falls, west of Thunder Bay, Ontario. The waterfall forms where the Kaministiquia River flows over the Gunflint Chert.

3 Flint Island, near Schreiber, north shore of Lake Superior, Ontario.

from the middle Devonian of Scotland, UK (see Selden and Nudds, 2004). This was effectively the first documentation of well-preserved fossils deep in the Precambrian, but the paper was greeted with skepticism from both geologists and botanists alike. In the middle of the twentieth century, paleontologists were firmly rooted in the opinion that the Precambrian was barren of such fossils, and the Gunflint forms were dismissed as being geofacts.

A decade later, after Tyler's death, Barghoorn published a more comprehensive paper on the Gunflint biota in which he eventually applied Linnaean nomenclature to eight of the most common forms, such as *Gunflintia, Kakabekia,* and *Eoastrion* (Barghoorn and Tyler, 1965). This description of actual genera caught the imagination of both the scientific and popular press and was the catalyst for a blossoming of research programs through the 1960s.

Stratigraphic setting and taphonomy of the Gunflint Chert

The Gunflint Formation was originally named by Van Hise and Clements (1901) for iron-rich sediments in the Thunder Bay region of Ontario, and later changed to the Gunflint Iron Formation by Leith *et al.* (1935). The name comes from the former use of the Gunflint cherts as flints for early muzzle-loaders, their high silica content making them ideal for producing sparks.

The Formation has recently been dated at 1,878 million years and thus falls in the middle of the Paleoproterozoic Era (which together with the Mesoproterozoic and Neoproterozoic eras make up the Proterozoic Eon which extends from the base of the Cambrian, at 542 million years, back to 2,500 million years; Table 1). It outcrops almost continuously from Thunder Bay on Lake Superior to Gunflint Lake on the Minnesota–Ontario border and continues into Minnesota, a total distance of approximately 180 km (110 miles), and averages 122 m (400 ft) in thickness (**1**).

Together with the overlying and conformable Rove Shale Formation, the Gunflint Iron Formation constitutes the Animikie Group. These rocks are relatively undeformed and exhibit low-grade metamorphism, ranging from diagenetic to subgreenschist. Goodwin (1956) divided the Gunflint Iron Formation into four members (**4**), the Basal Conglomerate Member (or Kakabeka Conglomerate), the

Member	Facies	Thickness (feet)
Upper Limestone		5–20
Upper Gunflint	taconite tuffaceous shale algal chert and jasper	150–180 5–16 Total 200–280 48–86
Lower Gunflint	taconite tuffaceous shale algal chert	150–210 4–22 Total 210–240 2–15
Basal Conglomerate		1–5

4 Diagram to show the stratigraphy of the Gunflint Iron Formation (after Goodwin, 1956).

Lower and Upper Gunflint members, and the Upper Limestone Member. The Lower and Upper Gunflint members both exhibit a similar sedimentary cyclicity going in ascending order from algal chert, to shale, to chert-taconite, and finally to chert-carbonate. The Lower Gunflint Member thus begins with the Lower Algal Chert Facies, and it is this unit that has yielded most of the microbiota, and in many locations is characterized by stromatolitic bioherms (**5, 6, 7**).

Taphonomically, the Gunflint cherts and stromatolites differ from those normally encountered in the Precambrian. Most cherts with microfossils occur as lenses or as thin beds which formed *within* carbonate sediments after deposition. The Gunflint cherts, however, are thought by many to have formed by silica precipitation directly onto the sea floor. Likewise, stromatolites are normally found in carbonates, but at Gunflint they actually occur in the cherts; Knoll (2003) suggested that the Gunflint 'stromatolites' may actually be sinters – laminated build-ups of silica precipitated from SiO_2 springs as in Yellowstone Park today and not necessarily produced by microorganisms.

Large populations of microfossils lay jumbled and disorientated on successive laminae of the stromatolitic structures, but rather than these being the organisms that built the laminae, perhaps they merely fell onto the accreting surfaces to be entombed and preserved in silica, just as in the Devonian Rhynie Chert of Scotland in the UK (see Selden and Nudds, 2004). Other scientists do not agree that the chert is a primary silica chemical sediment and have presented evidence that the initial mineral was a carbonate, possibly aragonite, which was later silicified (Sommers *et al.*, 2000).

Description of the Gunflint Chert Biota

The first eight genera described here correspond to those originally described by Barghoorn and Tyler (1965), while the last one was described subsequently by Knoll *et al.* (1978); these are the most common and best known of the Gunflint fossils. Several other taxa have been described (see Awramik and Barghoorn, 1977), but most are either of doubtful biological origin or are very rare and are not included here.

5 Domed stromatolitic bioherm growing on the Archean erosional surface, near Kakabeka Falls Provincial Park, Ontario.

6 Circular stromatolitic bioherm in Gunflint Chert, on the north shore of Lake Superior opposite Flint Island, near Schreiber, Ontario.

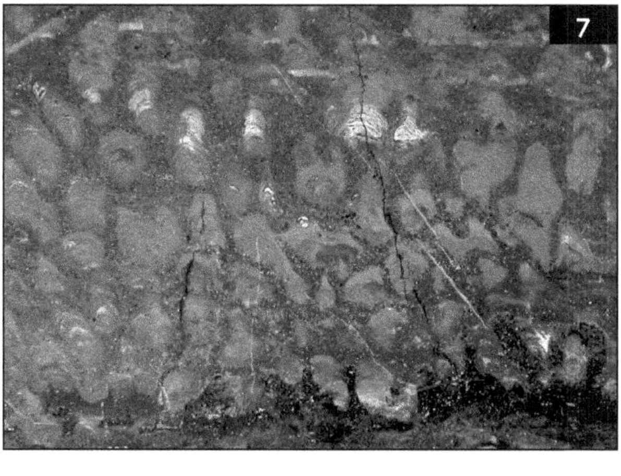

7 'Mary Ellen Jasper', a stromatolitic red chert from the Mesabi Iron Range of Minnesota, equivalent to the Gunflint Chert (MM).

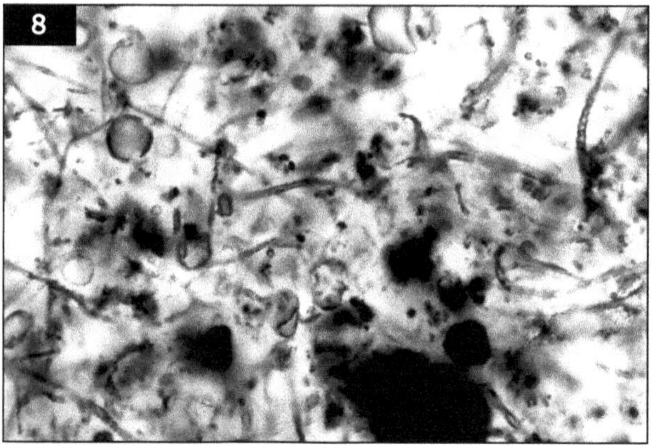

8 The filamentous bacteria *Gunflintia* (with spherical bacteria *Huroniospora*) (PC). Filaments 1–5 micron diameter.

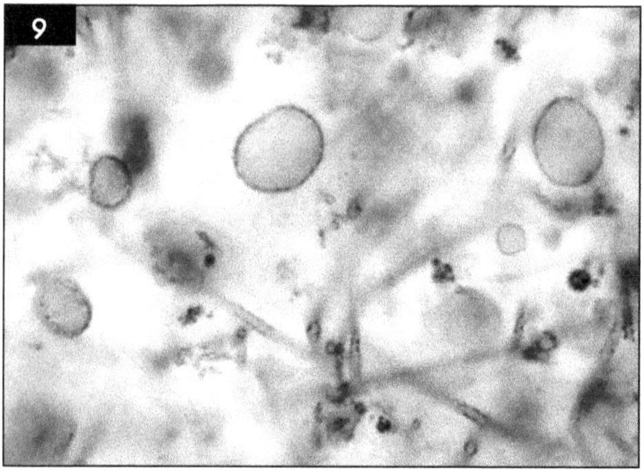

9 The spherical bacteria *Huroniospora* (with filamentous bacteria *Gunflintia*) (PC). Spheres 1–16 micron diameter.

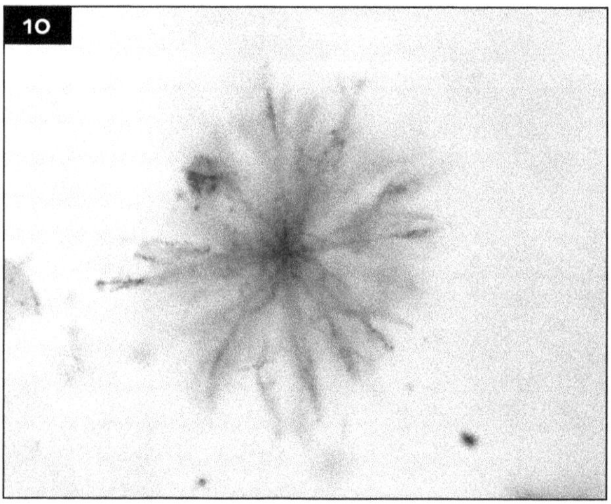

10 The star-shaped bacteria *Eoastrion* (PC). Filaments 1.5 micron diameter.

Gunflintia and other filaments

Several different genera have been described of tiny filamentous forms. In *Gunflintia*, which is the most common Gunflint fossil, the filaments have thickened partitions (septa) spaced at irregular intervals so that the individual cells are of variable size. The cells are longer than they are wide, 1–5 microns in diameter, and the filaments are often over 300 microns long (**8, 9**). *Animikiea* has closely spaced septa so that individual cells are wider than they are long, and are from 7–10 microns in diameter. These filaments may be up to 100 microns long and the sheath enclosing the filaments has a granular appearance. The filaments of *Entosphaeroides* have no dividing septa; they are 5–6 microns in diameter and up to 100 microns long. Finally, *Archaeorestis* differs from the previous three in that its filaments are branched. It has no dividing septa, and possesses characteristic rugose walls. A bulbous swelling of the filament wall sometimes occurs randomly; filament diameter ranges from 2–10 microns and filaments are up to 200 microns long. All these filamentous fossils resemble modern thread-like bacteria, including both blue-green cyanobacteria and siderophile (iron-loving) bacteria living in iron-rich waters today.

Huroniospora

These tiny fossils are spheroidal to ellipsoidal unattached bodies with a long axis 1–16 microns long. The thick wall is sculptured with a reticulate pattern and they may have a minute aperture at the restricted end (**8, 9**). They resemble coccoidal iron-loving bacteria.

Eoastrion

This is one of the best known of the Gunflint fossils. A variable number of septate filaments radiate from a central body and form a characteristic star shape. Filaments may be single or branched and are 1.5 microns in diameter, but only 3–18 microns long (**10**). These 'little dawn stars' can be compared with extant bacteria that oxidize iron and manganese in their metabolism.

Kakabekia

This lovely little fossil has a distinctive threefold structure consisting of a spheroidal bulb, a slender stipe, and an umbrella-like crown. The overall length is about 12–30 microns (**11, 12**) and

11 The umbrella-shaped bacteria *Kakabekia* (PC). Length 12–30 micron.

12 Reconstruction of *Kakabekia*.

examples are extremely abundant from the Kakabeka Falls locality, hence their name. Amazingly some extant forms of this genus have since been discovered in modern soil samples from the latrines of Harlech Castle, North Wales, where they live in reducing, ammonia-rich environments (Siegel and Giumarro, 1966). Where were they during the intervening 2 billion years?

Eosphaera

This is another very distinctive fossil, consisting of a sphere within a sphere. On the outer surface of the thick-walled inner sphere are rugose spheroidal tubercles. The radius of the inner sphere is from 10–12 microns (**13**).

Leptoteichos

This is a larger spherical microfossil (up to 31 microns diameter) usually occurring singly, but sometimes in clumps of up to 100 individuals. The most important feature is that many specimens contain an internal dark spot (**14**), which some workers have been tempted to interpret as a cell nucleus and hence to suggest that this genus is a very early eukaryote. However, much more evidence would be required to confirm such a hypothesis; all other Gunflint microbes are prokaryotic. Similar spherical unicells have been found in the coeval Duck Creek Formation of Western Australia (Knoll and Barghoorn, 1975).

Paleoecology of the Gunflint Chert

The Gunflint Iron Formation represents a Paleoproterozoic assemblage of chemical and siliciclastic sedimentary rocks deposited in a shallow water, south-facing shelf setting on the southern margin of Superior Province, Ontario, Canada. The occurrence of near-shore tidal flat sequences in the Gunflint requires unrestricted connection with a large ocean basin. Correlative units to the Gunflint Formation outcrop over several hundred kilometres to the southwest (e.g. the Biwabik Formation of Minnesota), and represent progressively deeper-water deposits with no evidence of basin closure. Therefore, deposition of the Gunflint Formation most probably occurred in a shelf-slope setting with open connection to the ocean, although some authorities have suggested that it was a large lake.

The Gunflint is a classic example of a Banded Ironstone Formation or 'BIF' – a sedimentary deposit of iron oxides

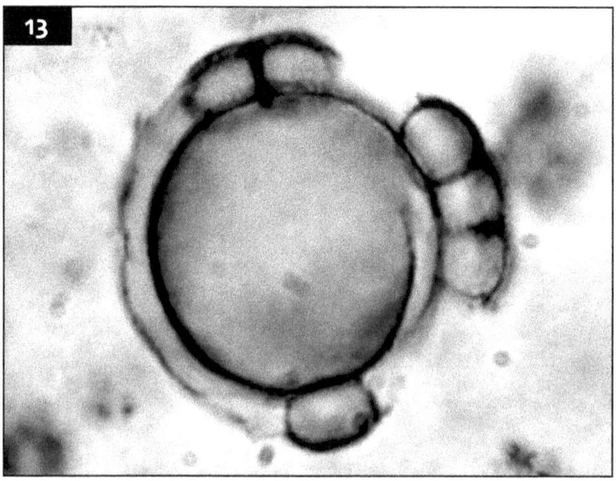

13 The double-sphered bacteria *Eosphaera* (PC). Radius of the inner sphere 10–12 micron.

(mostly hematite) alternating with bands of silica, often in the form of chert. BIFs are characteristic of the Gondwana continents and (with only a few exceptions) are restricted to quite a narrow period of time within the early Precambrian. There is a definite peak of BIFs at about 2.5 billion years, but there is also quite a sharp line defining their sudden increase (at about 2.8 billion years) and dramatic demise (a little after 2 billion years). These peculiar formations, which are almost never seen again in the geological record, tell us something about the composition of Earth's early atmosphere and oceans.

Earth's early atmosphere was almost devoid of oxygen (well under 1%). The first bacteria to inhabit the planet were thus anaerobic (i.e., they were able to live in the absence of oxygen); in fact, oxygen would have been lethal to them. However, some types of bacteria (including the cyanobacteria) were photosynthesizing and constantly producing oxygen as a waste product. In order for them to survive there had to be some mechanism which would rid the system of this poison. One direct result of the paucity of oxygen in the atmosphere was that the Earth's early oceans were rich in dissolved iron in its unoxidized ferrous state. This was the key to the disposal of the oxygen produced by photosynthesis. Oxygen is a very reactive element and as soon as it was released by photosynthesis it reacted with the ferrous iron to form iron oxide, which being insoluble was deposited on the ocean floor – the oceans were turning to rust!

At first this only happened slowly – the rate at which the photosynthesizing bacteria could increase was limited by the speed at which their waste oxygen could be processed into ferric iron. However, at some point in time the photosynthesizing bacteria evolved to become aerobic – suddenly they could live happily in the presence of oxygen and there was no longer any restriction on their increase. A population explosion of photosynthesizing bacteria occurred, producing huge amounts of oxygen, which flushed the ferrous iron out of solution, producing vast amounts of iron oxide deposited on the seabed; hence the sudden increase in BIFs around 2.8 billion years ago. BIFs persisted until all the available ferrous iron dissolved in the oceans had been used up (sometime after 2 billion years ago), explaining their equally sudden

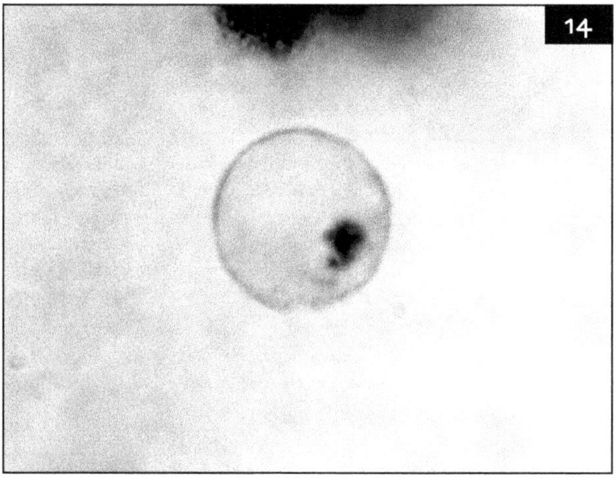

14 The large, spherical bacteria *Leptoteichos* with possible internal nucleus. (PC). Up to 31 micron diameter.

demise. At this point all of the oxygen produced by photosynthesis was now able to build up gradually in the atmosphere, eventually reaching the present day level of 20% (with great significance for the later evolution of multicellular forms; Chapter 2).

This elegant theory was put forward by Preston Cloud in 1965 and it is still generally accepted, although other factors are also now known to be involved in delaying the eventual build-up of oxygen in the atmosphere (see Knoll, 2003 for discussion). Nonetheless, the Precambrian BIFs tell us that Earth's early atmosphere was low in oxygen while its oceans were rich in dissolved iron – in other words the Gunflint seaway was unusual and not at all similar to any habitat in any modern ocean. (These early oceans may also have been much warmer, but the maximum temperature tolerated by photosynthesizing organisms is about 74°C.)

These unusual marine conditions are reflected in the biota; the anaerobic Gunflint ecosystems may have consisted of nothing but minute bacteria, but there was a wonderful variety of these microorganisms. While eukaryotes (protists, fungi, plants, and animals) use either respiration, photosynthesis, or fermentation to metabolize energy, the prokaryotic bacteria use all of these methods plus chemosynthesis, where energy is harvested from chemical reactions. Oxygen or nitrate or even sulfate, iron oxide, or manganese oxide, is combined with hydrogen, methane, or the reduced forms of iron, sulfur, and nitrogen and the cells capture the energy released by the reaction. Moreover, some respiring bacteria use nitrate, sulfate, iron oxide, or manganese oxide instead of oxygen.

It is still uncertain whether the Gunflint stromatolites represent true microbial mat structures built by oxygen-generating cyanobacteria (Lanier, 1989) or whether they are abiogenic sinters (Knoll, 2003). If they are true stromatolites, then we can presume that the filamentous forms were autochthonous, growing *in situ* and contributing to the mat-building process, while the spheroidal, star-shaped and umbrella-like types were planktonic and allochthonous.

Whichever scenario is correct, we should assume that the Gunflint biota included some photosynthesizing cyanobacteria (e.g. the larger filaments and enveloped coccoids), but undoubtedly many do appear to have used metabolic strategies suited to the peculiar chemical conditions of the Gunflint sea. For example, the iron-coated tubes of *Gunflintia minuta,* the most common fossils, resemble the sheaths of iron-loving bacteria living in iron-rich waters today, while the branching tubes of *Archaeorestis* resemble the iron-oxidizing bacteria *Gallionella ferruginea,* which presently inhabits hydrothermal vents of the Arctic Ocean. Similarly, the tiny, spherical *Huroniospora* are similar to coccoidal iron-loving bacteria, and the 'little dawn stars' (*Eoastrion*) can be compared with living bacteria that use both iron and manganese in their metabolism (Knoll, 2003). Finally, *Kakabekia* persists today in microhabitats where oxygen is reduced and ammoniacal decomposition takes place. These nonphotosynthesizing microbes probably included both iron oxidizing chemosynthesizers (which are autotrophic) and iron respirers (which are heterotrophic), and compare to a rare set of modern counterparts that are not much evident in today's iron-starved oceans.

The holistic view of Proterozoic life concluded by Knoll (2003) is that true microbial mat stromatolites were abundant in carbonate-rich environments throughout all of the Proterozoic and during much of the preceding Archean Eon, and that the oxygen-generating cyanobacteria which built them were the primary producers in early oceans. Occasionally, however, an upwelling of anoxic deep ocean waters containing dissolved iron mixed with oxygenated surface waters on the shelf-slope, and Gunflint-type iron-loving bacteria flourished. Such Gunflint-type microbiotas have since been recorded from coeval formations in other parts of the world (for example, the Duck Creek Dolomite of Western Australia; Knoll and Barghoorn, 1975), suggesting that they were cosmopolitan and representative of the level of evolution that had been reached 2 billion years ago, rather than being just a local phenomenon.

However, soon after this time the Gunflint-type bacteria almost became extinct, not due to competition with cyanobacteria, but due simply to the loss of their habitat once the ferrous iron had been flushed from the early oceans by the process of oxidization. Never again would organisms that neither use nor produce oxygen be able to dominate Earth's ecosystems in the manner of the Gunflint microbes.

Comparison of the Gunflint Chert with other Precambrian microbial biotas

North Pole, Western Australia

The 3,500 million year old Apex Chert of the Warrawoona Group, which outcrops in the Marble Bar region of Western Australia, contains the earliest fossil evidence of life on this planet. Domed stromatolites were first discovered at Warrawoona in 1980 by Don Lowe (of Stanford University) and by Roger Buick and his colleagues (then of Harvard); then in 1987 more convincing evidence was found when Bill Schopf of UCLA discovered tiny microfossils in the cherts of Chinaman's Creek, near North Pole. These fossils, 1–20 microns in diameter and a few hundred microns long, looked like simple cyanobacterial filaments (Schopf and Packer, 1987; Schopf, 1993) and for a decade these were accepted as the earliest evidence of life on Earth.

However, doubt has since been cast on both the stromatolites and microfossils. The stromatolites have been reinterpreted as chemically deposited layers, which have been deformed tectonically. Moreover, Martin Brasier of Oxford University has recently cast doubt on the microfossils (Brasier *et al.*, 2002), suggesting that they are merely chains of crystals formed in a hydrothermal vein. Schopf still insists that the structures contain organic matter, but evidence of life at Warrawoona must remain doubtful unless further, more convincing microfossils are found from these rocks. However, as Knoll (2003) points out, carbon and sulfur isotopes do suggest that photosynthesizing and sulfate-reducing bacteria *did* live in the Warrawoona lagoon.

Barberton Greenstone Belt, near Kruger Park, South Africa

This is the only other place in the world where rocks approaching 3,500 million years old show fossil evidence of early life. The Fig Tree Chert (dated between 3,200 and 3,400 million years old) contains stromatolites of uncertain origin, and spherical and filamentous microstructures, 2–4 microns in diameter within cherts. This is the correct size for cyanobacteria, the structures are made of organic matter and possess an outer wall. However, nonbiological processes can produce similar structures and, like Warrawoona, these presently remain in doubt.

Somerset Island, Arctic Canada

The cherts of the Hunting Formation of Somerset Island, arctic Canada are much younger than Gunflint, dated at 1,200 million years old. These stratified cherts within intertidal to supratidal carbonates contain fossils of a well-preserved bangiophyte red alga (seaweed) belonging to the extant genus *Bangia*. This is the first undisputed eukaryote to be recorded in the fossil record and shows us that multicellular algae diversified well before the Ediacaran radiation of animals (Chapter 2). Found with the cyanobacterium, *Polybessurus*, and with stratiform stromatolites, the algae have a mean diameter of 26 microns and form unbranched filaments with their basal cells differentiated into holdfasts attached to the substrate (Butterfield *et al.*, 1990).

Spitsbergen

Younger still, the 700–800 million year old chert nodules of the Draken Conglomerate Formation (Akademikerbreen Group) of northeastern Spitsbergen contain a diverse microbiota of over 40 taxa from a tidal flat/lagoon complex associated with stratiform stromatolites. Many of the microfossils compare very closely to modern day cyanobacteria (e.g., *Polybessurus bipartitus*), but preserved with them are some larger (>100 microns) vase-shaped forms and others studded with spines; these constitute the first really diverse biota of eukaryotes (Knoll *et al.*, 1991).

Doushantuo Formation, southern China

Discovered in the 1980s by Zhang Yun of Beijing University, are beautiful microfossils preserved in cherts and shales which outcrop in the Yangtze Gorge of Guizhou Province, southern China. These rocks are only c.600 million years old, and were deposited a mere 80 million years before those of the Burgess Shale (Chapter 3). They include a wealth of eukaryotic fossils including red algae (the corallines), plus an amazing fauna discovered in 1995 of tiny spheres, 500 microns in diameter, which exhibit cell division into pairs, quartets, octads, and so on, and have been interpreted as the eggs and embryos of animals – the 'Cambrian Explosion' was yet to happen, but animal evolution had at last begun.

Further Reading

Awramik, S. M. and Barghoorn, E. S. 1977. The Gunflint microbiota. *Precambrian Research* **5**, 121–142.

Barghoorn, E. S. and Tyler, S. M. 1965. Microfossils from the Gunflint chert. *Science* **147**, 563–577.

Brasier, M. D., Green, O. R. and Jephcoat, A. P. *et al.* 2002. Questioning the evidence for Earth's oldest fossils. *Nature* **416**, 76–81.

Butterfield, N. J., Knoll, A. H. and Swett, K. 1990. A bangiophyte red alga from the Proterozoic of Arctic Canada. *Science* **250**, 104–107.

Cloud, P. 1965. The significance of the Gunflint (Precambrian microflora). *Science* **148**, 27–35.

Goodwin, A. M. 1956. Facies relations in the Gunflint Iron Formation. *Economic Geology* **51**, 565–595.

Knoll, A. H. 2003. *Life on a Young Planet.* Princeton University Press, Princeton.

Knoll, A. H. and Barghoorn, E. S. 1975. A Gunflint-type flora from the Duck Creek Dolomite, Western Australia. *Abstracts of the College Park, Maryland, Colloquia on Chemical evolution of the Precambrian*, p. 61.

Knoll, A. H., Barghoorn, E. S. and Awramik, S. M. 1978. New microorganisms from the Aphebian Gunflint Iron Formation, Ontario. *Journal of Paleontology* **52**, 976–992.

Knoll, A. H., Swett, K. and Mark, J. 1991. Paleobiology of a Neoproterozoic tidal flat/lagoon complex: the Draken Conglomerate Formation, Spitsbergen. *Journal of Paleontology* **65**, 531–570.

Lanier, W. P. 1989. Interstitial and peloidal microfossils from the 2.0 Ga Gunflint Formation: implications for the paleoecology of the Gunflint stromatolites. *Precambrian Research* **45**, 291–318.

Leith, C. K., Lund, R. J. and Leith, A. 1935. Pre-Cambrian rocks of the Lake Superior region, a review of newly discovered geologic features, with a revised geological map. *United States Geological Survey, Professional Paper* **184**.

Schopf, J. W. 1993. Microfossils of the early Archean Apex Chert: new evidence of the antiquity of life. *Science* **260**, 640–646.

Schopf, J. W. and Packer, B. M. 1987. Early Archean (3.3 billion to 3.5 billion-year-old) microfossils from Warrawoona Group, Australia. *Science* **237**, 70–73.

Selden, P. and Nudds, J. 2004. *Evolution of Fossil Ecosystems.* Manson, London.

Siegel, S. M. and Giumarro, C. 1966. On the culture of a microorganism similar to the Precambrian microfossil *Kakabekia umbellata* Barghoorn in NH-rich atmospheres. *Proceedings of the National Academy of Sciences of the United States of America* **55**, 349–353.

Sommers, M. G., Awramik, S. M. and Woo, K. S. 2000. Evidence for initial calcite-aragonite composition of Lower Algal Chert Member ooids and stromatolites, Paleoproterozoic Gunflint Formation, Ontario, Canada. *Canadian Journal of Earth Science* **37**, 1229–1243.

Tyler, S. A. and Barghoorn, E. S. 1954. Occurrence of structurally preserved plants in Precambrian rocks of the Canadian shield. *Science* **119**, 606–608.

Van Hise, C. R. and Clements, J. M. 1901. The iron-ore deposits of the Lake Superior region. 305–434 In: *United States Geological Survey, Annual Report, Washington, D.C.* C. R. Van Hise and C. K. Leith (eds.).

Chapter 2

Mistaken Point

Background: the first animals

The development of multicellularity was a major step in the evolution of life: it enabled organisms to grow in size, to develop organ systems through tissue differentiation, and led to the plants and animals we are familiar with today.

Until the middle of the last century it was thought that rocks older than Cambrian in age, collectively called Precambrian, were devoid of fossils of multicellular creatures. The base of the Cambrian is clearly marked by the sudden appearance of shelly fossils – brachiopods, mollusks, trilobites and sponges for example – but this posed a difficult question for Darwin whose theory of gradual evolution required a long period of pre-Cambrian diversification of these varied groups.

The discovery of soft-bodied organisms similar in appearance to jellyfish and worms in rocks of late Precambrian age came as a major surprise. As Andrew Knoll (2003) eloquently puts it, in some ways it marked the realization of Darwin's dream, i.e. animal life before the Cambrian, but it also deepened Darwin's dilemma, because the Precambrian forms seem to bear little resemblance to anything that came either before or after. These peculiar organisms were typically flat creatures, with a high surface area/body mass ratio, and have posed more questions than they have answered. How they relate to the typical Cambrian animals, and what happened to them at the end of the Precambrian are questions that will be explored in this chapter.

This assemblage of primitive organisms is often known as the 'Ediacaran fauna' after their initial discovery in the Ediacara Hills of South Australia (see Selden and Nudds, 2004), but similar fossils have since been found at approximately 30 localities on five continents, notably in the United Kingdom, Namibia, Ukraine, and in the White Sea area of Russia. One of the most informative sections, however, is the spectacular locality of Mistaken Point, a wave-swept crag on the southernmost tip of the Avalon Peninsula in southeast Newfoundland, Canada, which not only exposes the thickest Ediacaran successions anywhere in the world, but also exhibits complete life assemblages, sometimes of thousands of individuals, buried beneath beds of volcanic ash, such that one can walk across a sea floor apparently frozen in Ediacaran time. Here then is preserved a record of that critical period in the Earth's history, which falls between the microbial ecosystems of the Precambrian and the animal ecosystems of the Phanerozoic.

History of discovery of the Mistaken Point biota

The unusual and unexpected Ediacaran fossils had first been discovered in the Ediacara Hills of South Australia by government geologist, Reg Sprigg, in 1946 (Sprigg, 1947), though at the time it was assumed that they were Cambrian in age simply because the Precambrian was presumed to be devoid of macrofossils. Their true age was not established until 10 years later when an English schoolboy, Roger Mason, discovered similar fronds and disc-like fossils from unequivocal Precambrian strata in Charnwood Forest, England, later described by Trevor Ford of Leicester University (1958).

It was to be a further decade, however, in the summer of 1967, before similar primitive fossils were discovered at Mistaken Point by Shiva Balak Misra, an Indian graduate student from Newfoundland's Memorial University. Misra was studying an unmapped part of the Avalon Peninsula (**15**) when he came across a rich assemblage of soft-bodied organisms on the bedding

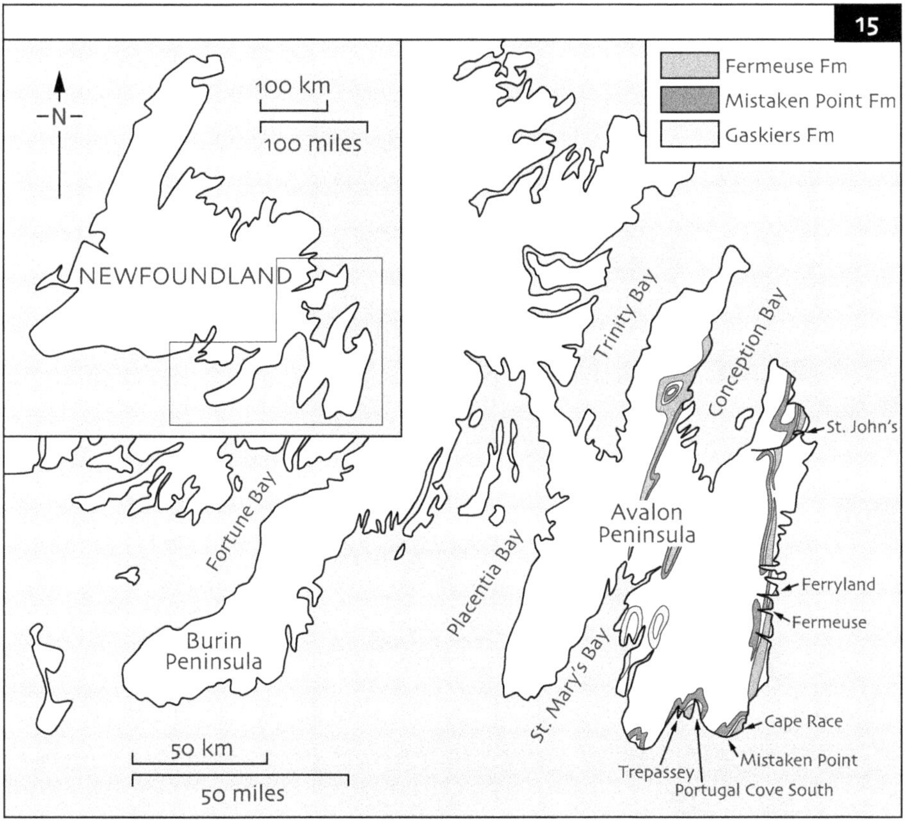

15 Locality map showing the Avalon Peninsula of southeast Newfoundland and outcrop of the main stratigraphical units (after Gehling *et al.*, 2000).

planes of mudstones exposed at Mistaken Point (**16, 17**). Some of these he recognized as comparable to the Australian forms, but most were unique and defied identification.

The discovery was immediately reported in *Nature* (Anderson and Misra, 1968), but, just as with the Gunflint biota described in Chapter 1, the organic nature of these fossils was initially doubted until subsequent reports documented the entire diverse assemblage (Misra, 1969). Misra classified the primitive animals into four groups, namely spindle-shaped, leaf-shaped, round-lobate, and dendrite-like, and believed that most of them represented polyp and medusoid forms of cnidarians.

Ironically, the most common of these fossils (*Aspidella terranovica*) had actually been described from Newfoundland almost 100 years previously (Billings, 1872) from beds 1 km (0.62 miles) higher in the succession, but had been dismissed by most authorities as an inorganic pseudo-fossil simply because it was Precambrian in age. As has now been pointed out by Gehling *et al.* (2000), *Aspidella* was actually the first named member of the Ediacaran biota from anywhere in the world.

16 Mistaken Point on the southernmost tip of the Avalon Peninsula, southeast Newfoundland.

17 Bedding planes of mudstones with soft-bodied fossils exposed at Mistaken Point.

Stratigraphic setting and taphonomy of the Mistaken Point biota

The late Neoproterozoic sediments and interbedded volcaniclastics exposed along the Avalon Peninsula in southeastern Newfoundland comprise a conformable succession more than 7.5 km (4.7 miles) thick. They thus constitute an outstanding record of this crucial period of geological time immediately following the severe global glaciation referred to as the 'Snowball Earth'.

This thick succession has been divided into three stratigraphic groups, further subdivided into several formations (**18**). Above the massive glacial deposits of the Gaskiers Formation, the Conception Group consists dominantly of thick-bedded, coarse-grained turbidites (greywackes and

18 Diagram to show the stratigraphy of the late Neoproterozoic succession of the Avalon Peninsula, southeast Newfoundland (after Gehling *et al.*, 2000; Narbonne *et al.*, 2001).

Cambrian	Stratigraphical Units		Fossil Assemblages
Terminal Proterozoic	Signal Hill Group	Cape Ballard Fm	
		Cuckold Fm	
		Ferryland Head Fm	
		Gibbett Hill Fm	
		Cappahayden Fm	
	St John's Group	Renews Head Fm	
		Fermeuse Fm	Fermeuse Assemblage
		Trepassey Fm	Mistaken Point Assemblage
	Conception Group	Mistaken Point Fm	
		Briscal Fm	
		Drook Fm	
		Gaskiers Fm	
		Mall Bay Fm	

siltstones), but the Mistaken Point Formation at the top of this group represents a transition to the thinner-bedded and finer-grained shales and sandstones of the overlying St John's Group. Although this succession is interpreted as upwardly shallowing, it is all considered to be a deep-water regime. The overlying Signal Hill Group consists mainly of deltaic and fluvial deposits and thus continues this upward shallowing trend.

Ediacaran-type fossils range from the upper part of the Drook Formation in the Conception Group, through to the top of the Fermeuse Formation in the St John's Group, and span some 3 km (1.9 miles) of strata (**18**). Fossils are known from over 100 different bedding surfaces and many of these horizons are overlain by thin volcanic tuffs, which have allowed accurate uranium/lead (U/Pb) radiometric dating of included zircon crystals. The most richly populated fossil bed, which occurs in the Mistaken Point Formation, has been dated at 565 million years, while an ash bed in the upper Drook Formation has yielded an age of 575 million years, making these the oldest accurately dated Ediacaran fossils in the world, and therefore the oldest known animal communities.

The presence of these tuffs in the Mistaken Point Formation also has an important bearing on the taphonomy of the fossils, which shows marked differences to the preservation of Ediacaran assemblages in other parts of the world where they are preserved on the soles of sandstone event beds (see Selden and Nudds, 2004, Figure 5). Here, on the other hand, specimens are preserved on the muddy tops of turbidite beds underneath volcanic tuffs, suggesting a smothering of benthic life assemblages by episodic ash falls. Not only did this ensure that populations were preserved intact, but the fine ash also preserved the detailed morphology of these soft-bodied organisms. Such 'death masks' in the overlying ash tend to mould the more resistant fronds and to cast the easily collapsed spindle-shaped organisms. Morphological detail is unfortunately limited by the grain size of the tuff and some forms can often only be described in terms of gross morphology. In the Fermeuse Formation, however, the fossils are associated with thin sandstone beds in the preservational style seen in Ediacaran assemblages in other parts of the world. Some unusual specimens have recently been discovered by Guy Narbonne (2004) in the Trepassey Formation, which are preserved as uncompressed external moulds and which exhibit micron-scale features including previously unseen internal structures (see p. 37).

Description of the Mistaken Point Biota

The diverse fauna of up to 30 species (most of which have yet to be given formal names) differs from Ediacaran faunas elsewhere in the world in that only a few of the taxa are cosmopolitan (e.g. *Aspidella*, *Charnia*, *Charniodiscus*). Others are known from equivalent strata at the classic locality of Charnwood Forest in England (e.g. *Bradgatia* and *Ivesia*), but most are endemic and restricted to the Mistaken Point assemblages. Moreover, the well-known and ubiquitous Ediacaran genera such as *Tribrachidium*, *Dickinsonia*, *Spriggina*, and *Kimberella* (see Selden and Nudds, 2004) are unknown from this locality. Two distinct assemblages occur: the diverse Mistaken Point Assemblage consists mainly of fronds and spindle-shaped forms, while the younger and less diverse Fermeuse Assemblage contains mostly small discoidal forms (**18**).

Aspidella

This prolific fossil ranges from the Briscal Formation to the top of the Fermeuse Formation (**18**) where it reaches its zenith literally in tens of thousands of individuals. It was the first fossil to be described from this assemblage (Billings, 1872) and is typically a small disc (5 mm [0.2 in] diameter), but shows immense variation in size (up to 100 mm [4 in] diameter) and preservational styles (Gehling *et al.*, 2000). Most commonly in negative hyporelief, it exhibits a raised rim and ridges radiating from a slit (**19**), while in positive hyporelief it has concentric ornamentation, or it may be a flat disc with a central boss. A few specimens possess a stalk and appear to be the holdfasts of fronds of *Charnia*. It may be that *Aspidella* includes the holdfasts of more than one genus of frond.

Ivesia

These larger discoidal impressions (5–60 cm [2–24 in] diameter) are also highly variable, including 'lobate discs' with inwardly pointing V-shaped ridges, 'bubble discs' with rounded craters near the margin, and the peculiar 'pizza discs' with closely-spaced pustules covering the surface (**20**). The presence of a stalk on the latter suggests that it may also be a holdfast or even a floating body tethered to the seabed (Narbonne *et al.*, 2001).

19 The disc-like fossil *Aspidella* from the Fermeuse Formation at Ferryland, Avalon Peninsula. Coin 22 mm (0.9 in) in diameter.

20 The disc-like fossil *Ivesia*, known as 'pizza discs', from the Drook Formation at Pigeon Cove, Avalon Peninsula. Scale bar is in cm and inches.

Charnia

At least two distinct species of *Charnia* are known. *C. masoni* is the same species of segmented frond originally described from Charnwood Forest in England by Trevor Ford. It has a leaf-shaped appearance with lateral branches (**21, 22**), is sometimes found attached to a discoidal holdfast, and is generally accepted as a sea pen, an extant group of cnidarians. A recently described new species, *C. wardi* (Narbonne and Gehling, 2003), from the Drook Formation, represents both the largest known Ediacaran fossil, up to nearly 2 m (6 ft) in length, and the oldest.

21 The frond-like *Charnia masoni* from the Mistaken Point Formation at Mistaken Point. Coin 30 mm (1.2 in) in diameter.

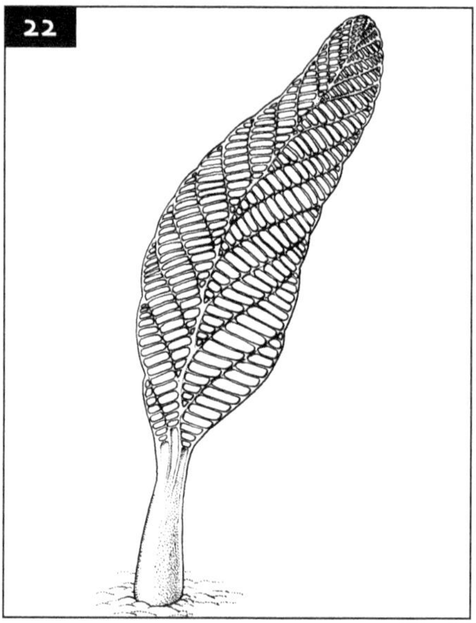

22 Reconstruction of *Charnia*.

Charniodiscus

This genus was originally described, also from England by Ford, as a discoidal form, but later discoveries showed that it was the holdfast of a *Charnia*-like frond, but which differed sufficiently in its branching pattern to be regarded as a distinct genus (Jenkins and Gehling, 1978). Some of those from Mistaken Point are similar to the well-known *Charniodiscus arboreus* from South Australia; like *Charnia*, it is regarded as a pennatulacean cnidarian (**23**).

Bradgatia

This is a bush-shaped form and varies from oval (4–12 cm [1.5–5 in] long) with the confluence of branching at one end, to circular (10–16 cm [4–6 in] diameter) with branching originating from within the structure (**24**). It is unclear whether it is a body fossil or a trace fossil.

Fractofusus

The most common member of the Mistaken Point Assemblage (up to 1000 individuals

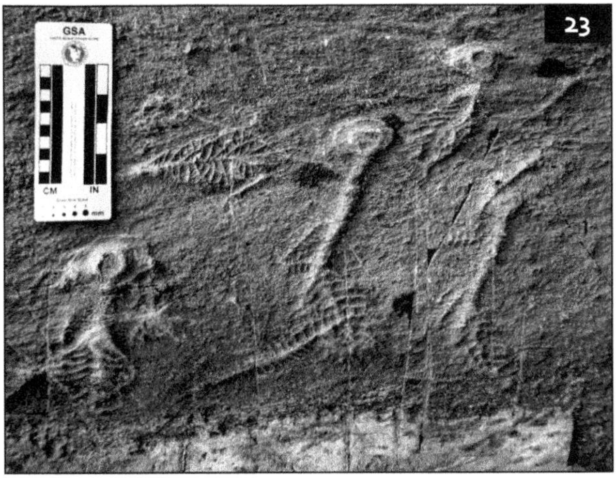

23 Four specimens of the frond-like *Charniodiscus* with circular holdfasts, with a specimen of *Charnia masoni* (to the right of the scale bar). From the Mistaken Point Formation at Mistaken Point. Scale bar is in cm and inches.

24 The bush-shaped fossil *Bradgatia* from the Mistaken Point Formation at Mistaken Point. Coin 30 mm (1.2 in) in diameter.

on a single bedding plane, usually cast by the volcanic ash) is an unusual spindle-shaped fossil consisting of a bipolar array of branches emanating from a zigzag midline (**25, 26**) and up to 30 cm (12 in) long. The branches may be opposite or alternating along the midline and there are at least two distinct forms. Their lifestyle was probably sessile, but their exact affinities are unknown. These characteristic fossils, so common in Newfoundland, have not been recorded in Ediacaran assemblages anywhere else in the world (Gehling and Narbonne, 2007).

Triforillonia

This interesting fossil, described by Gehling *et al.* (2000) from the Fermeuse

25 The spindle-shaped fossil, *Fractofusus* from the Mistaken Point Formation at Mistaken Point. Coin 30 mm (1.2 in) in diameter.

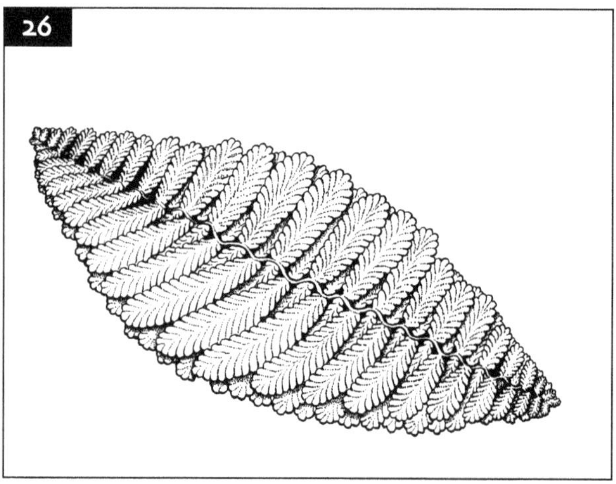

26 Reconstruction of *Fractofusus*.

Assemblage, possessed a three-lobed body with rounded lobes radiating from a central rosette (**27**). Threefold symmetry is rare in the animal kingdom, but does occur in another Ediacaran genus, *Tribrachidium*, first described from South Australia. *Triforillonia* may be a holdfast or a polyp-like organism, and was the first new genus to be named from the Newfoundland Ediacaran fauna since *Aspidella* in 1872; it is unknown elsewhere.

Thectardis

Another recently described endemic taxon (Clapham *et al.*, 2004), this triangular fossil has a raised rim and a faintly segmented central depression. Its authors consider that it was an elongate cone up to 15 cm (6 in) tall, that may have lived as a suspension feeding 'mat-sticker' with its pointed base inserted into the microbial bound sediment.

Others

Other unnamed frond-like fossils are colloquially known as 'ostrich feathers', 'feather dusters' and 'xmas trees' and await formal description. Rare pectinate, network, triangular and radial forms, and more common serial forms of curved sausage-shaped or bead-like elements, are similar to some described from Russia and Ukraine and recently documented from South Australia and Namibia. Some of the serial forms were originally interpreted as burrows, but are now thought by Jim Gehling to be impressions of soft-bodied bacterial colonies, algae, or egg masses growing on the substrate.

First attempts to classify the Ediacaran biota, by Sprigg (1947, 1949), Glaessner (1961), and Glaessner and Wade (1966), suggested a variety of extant phyla, mainly the Cnidaria and Annelida, but it was apparent that some Ediacaran forms could not be placed in any modern phylum. Later, Seilacher (1989) removed the Ediacaran organisms from the Metazoa altogether, preferring instead to regard them as a separate kingdom (Vendozoa) based on different functional design from plants and animals. He looked at the constructional morphology of the organisms and suggested they may have had internal hydrostatic skeletons, their bodies gaining rigidity through internal pressure, as in a car tyre. Compartmentalization of the body

27 The trilobate fossil *Triforillonia* from the Fermeuse Formation at Ferryland, Avalon Peninsula (GSC). Long axis 25 mm (1 in).

made sense to prevent total loss of pressure (and hence death) when only one or two segments were punctured. They were quilted organisms, similar to inflatable mattresses. The high surface area to mass ratio suggests that they respired through the skin; they may have been photosynthetic, as suggested by McMenamin (1998), used photosynthetic symbionts like modern corals, perhaps were chemosymbiotic (with chemosynthetic symbionts so they could survive in deep-water, reducing environments), or may have ingested materials through the body wall. Seilacher suggested that such organisms could survive without bony skeletons or shells because there was little predation, but that the kingdom became extinct with the onset of predation at the beginning of the Cambrian.

In the 1980s, Fedonkin devised a classification scheme for the Ediacaran fossils based on their symmetry and thus independent of any biological interpretation (1990). He divided the organisms into two major groups: Radiata (disc-shaped, no bilateral symmetry) and Bilateria (bilateral symmetry apparent). Radiata can be further subdivided into: Cyclozoa, with a concentric pattern (e.g. *Aspidella*); Inordozoa, with a radial pattern; and Trilobozoa with a three-rayed symmetry (e.g. *Triforillonia*). Other radiate forms were considered to belong to the cnidarian classes Conulata or Scyphozoa. Bilateria may show no particular head or tail (i.e. bidirectional, e.g. spindle-shaped forms), or have a distinct head or rooting structure (i.e. unidirectional, e.g. frondose forms). Intermediates exist, for example, between concentric and radial forms of *Aspidella*. One problem with this scheme, as pointed out by Jim Gehling, is that it only works if the organisms are flattened, yet many reconstructions show the creatures as quite three-dimensional.

In a strange twist to an already bizarre story, Greg Retallack of the University of Oregon looked at the indeterminate growth pattern of many of the Ediacaran organisms, which suggested that they had no upper limits nor definite bounds to their size and shape (Retallack, 1994). He also studied taphonomic aspects of the Ediacaran biota, particularly their compression. The conclusion? That Ediacaran fossils were lichens! Lichens are composite organisms formed by a symbiotic relationship between green algae (which provide photosynthesis) and fungi (which provide bulk). Few paleontologists have accepted the arguments presented by Retallack. While some Ediacaran organisms show a distinct holdfast (or head), and their growth patterns may be less determinate than worms, for example, they are not random; and the sheer size and bulk of many forms is quite unlike that of modern lichens. Moreover, modern lichens are terrestrial, not marine.

So what is the consensus? We can perhaps go someway towards a reconciliation of these apparently disparate theories by taking a slight retreat from Seilacher's (1989) view that they represent a separate kingdom, and considering instead that they represent a separate phylum (Buss and Seilacher, 1994), an animalian grade of organization more developed than simple sponges, in which few cell types are present, but far less evolved than the higher Eumetazoa in which tissues and organs have developed (see also Dewel *et al.*, 2001). Many of them seem to be colonial and similar in many respects to cnidarians (jellyfish, anemones, and corals), although not possessing all of the defining cnidarian features (such as a mouth surrounded by a ring of tentacles). Perhaps, therefore, they are on an evolutionary pathway towards the cnidarians (which are, after all, the simplest forms of eumetazoans), but have not yet attained the true cnidarian state. In this case we can consider them (or at least most of them) to belong to a single clade, which is at an early phase of the development of the metazoan body plans that we know today. Classifying these as a 'cnidarian-like' phylum of animals is not too removed from Glaessner's 1960s ideas that many of them were jellyfish or sea-pens, and that the Ediacaran biota represented 'the exposed roots of the metazoan tree of life'. Those few forms which do not fit into such a phylum probably can be accommodated in extant phyla, as has been shown by Jim Gehling who has identified early sponges, echinoderms, and mollusks in addition to the cnidarian-like forms (see page 38).

Any view expressed about the Ediacaran animals, however, will always find its opponents, and the microscopic architecture recently described in small frondose fossils by Narbonne (2004), including tertiary branches only 150 microns in diameter (far too small to accommodate cnidarian polyps), have led him back to Seilacher's original view that certainly the fronds and spindles (and maybe other forms) represent a clade (the 'rangeomorphs') that is unrelated to any modern group of organisms. Perhaps this older clade became extinct in the terminal Proterozoic in competition with the early 'cnidarian-like' animals represented in the younger Ediacaran assemblages both in Newfoundland, Australia, and elsewhere; if so, it represents a failed experiment in the evolution of multicellularity.

What happened to the Ediacaran biota after the end of the Precambrian? Some forms apparently continued into the Cambrian. For example the frond-like sea pen, *Thaumaptilon*, from the Burgess Shale (Chapter 3), is believed by Simon Conway Morris (1993) to be related to *Charniodiscus*. Moreover, Ediacaran-like fossils have been described from strata as young as the Upper Cambrian from County Wexford in Ireland (Crimes *et al.*, 1995). However, there is little doubt that the majority of Ediacarans did not survive to the Phanerozoic. Whether this was the result of the first mass extinction event, or whether the Ediacarans were simply eaten or outcompeted by Cambrian animals, or whether an increase in Cambrian bioturbation simply closed this taphonomic window, will continue to be debated for some time.

Paleoecology of the Mistaken Point biota

The Avalonian Terrain, which forms present-day eastern Newfoundland, together with England, Wales and parts of Northern Europe, and which was cleaved apart in the Mesozoic by the opening of the Atlantic Ocean, is thought to have been located adjacent to the Amazonian Craton during the late Neoproterozoic at about 40–65°S (Wood *et al.*, 2003). The presence of abundant ash layers in the successions suggests deposition associated with an arc-related basin.

Although Ediacaran assemblages from other parts of the world have been interpreted as marine benthic communities living in the shallow euphotic zone between fair-weather wave base and storm wave base, those from Mistaken Point seem to have been living as a deep-water slope biota, well below storm wave base (i.e., more than 50 m), on a southeast-facing slope. This interpretation is based both on the absence of any wave-generated features and by the fact that the succession is dominated by turbidites (graded beds deposited by downslope currents, with coarse sands and greywackes fining upwards into a muddy top). Most workers have previously suggested a submarine fan setting, but Wood *et al.* (2003) argue that the Drook and Briscal formations accumulated in a basin-floor axial turbidite system, while the finer-grained Mistaken Point, Trepassey, and Fermeuse formations formed on the marginal slopes of such a system.

The lack of evidence of uprooting of the tethered fronds or of damage to other soft-bodied forms, as might be expected in a turbidity flow, suggests that this assemblage lived, died, and was preserved *in situ* near the base of the slope at a depth of several hundreds of metres. Such an untransported census population of some thousands of individuals provides an ideal opportunity to study the ecology of this autochthonous community; Clapham *et al.* (2003) used techniques routinely applied to modern ecosystems, including parameters of species richness, organism abundance and biomass, and diversity and evenness coefficients.

Bedding planes at Mistaken Point, which record snapshots of the living communities at the moment that they were smothered by the volcanic ash, preserve an average organism density of 60 fossils per square metre (36 fossils/m^2 [4/ft^2] after retrodeformation; **23**), a species richness between 3–12 taxa per locality, and a biomass (expressed as a percentage of total surface area) ranging from 3.4% to 12.4%. Spindles comprise 30%, frondose and other upright forms (*Charniodiscus*, *Charnia*, and *Thectardis*) comprise 55%, while the remaining 15% includes *Bradgatia*, *Ivesia*, and *Hiemalora*. These values, combined with estimated diversity and evenness coefficients, all fall within the range of modern slope communities and suggest that ecological processes were remarkably similar in the Precambrian to those that operate today.

Comparison of Mistaken Point with other late Precambrian biotas

Ediacara Hills, South Australia

The classic locality in the Flinders Ranges, 300 km (190 miles) north of Adelaide in South Australia (see Selden and Nudds, 2004), is the most famous of the various Ediacaran assemblages now known from around the globe. Although it has yet to be dated accurately, it is possibly up to 10 million years younger than the Mistaken Point occurrence, and shows obvious differences in the composition of the fauna (as well as in its taphonomy and paleoecological setting – see previous sections). Because many of the Newfoundland species are endemic, it is really only the frond-like forms, such as *Charnia*, *Charniodiscus*, and their holdfast, *Aspidella*, which are common to both areas.

The South Australian assemblage is dominated by a variety of simple disc-like genera such as *Ediacaria* and *Cyclomedusa* (some of which may be synonyms of *Aspidella*), and also more complex discs such as *Mawsonites*, with concentric rows of lobate shapes getting larger from the center outwards. Another enigmatic disc is the trilobed *Tribrachidium*, which shares its unusual threefold symmetry with *Triforillonia* from Mistaken Point.

Perhaps best known from South Australia are the segmented forms such as *Dickinsonia* and *Spriggina*. Initially interpreted as flatworms and polychaete worms respectively, it was then pointed out that their segments do not actually match up across the midline and that they were therefore only 'pseudosegmented'. However, Jim Gehling, of the South Australian Museum, has countered this by showing that the segments do match perfectly on the rarely preserved ventral surfaces and suggests that the mismatch on dorsal surfaces is simply a result of squashing a three-dimensional animal.

Other important forms from South Australia include *Kimberella*, with fourfold symmetry, now interpreted as an early mollusk, *Arkarua*, with fivefold symmetry, thought to be the earliest known echinoderm, and *Paleophragmodictya*, described as the earliest known sponge. Another enigma is *Parvancorina*, thought by some authorities to be an early arthropod. The impression is that the slightly younger Australian forms are a bit more advanced than the fronds and spindles that dominate assemblages at Newfoundland, and have more in common with some of the Phanerozoic animals which were soon to replace them.

Further Reading

Anderson, M. M. and Misra, S. B. 1968. Fossils found in pre-Cambrian Conception Group of Southeastern Newfoundland. *Nature* **220**, 680–681.

Billings, E. 1872. Fossils in Huronian rocks. *Canadian Naturalist and Quarterly Journal of Science* **6**, 478.

Buss, L. W. and Seilacher, A. 1994. The phylum Vendobionta: a sister group of the Eumetazoa? *Paleobiology* **20**, 1–4.

Clapham, M. E., Narbonne, G. M. and Gehling, J. G. 2003. Paleoecology of the oldest known animal communities: Ediacaran assemblages at Mistaken Point, Newfoundland. *Paleobiology* **29**, 527–544.

Clapham, M. E., Narbonne, G. M., Gehling, J. G., Greentree, C. and Anderson, M. M. 2004. *Thectardis avalonensis*: a new Ediacaran fossil from the Mistaken Point biota, Newfoundland. *Journal of Paleontology* **78**, 1031–1036.

Conway Morris, S. 1993. Ediacaran-like fossils in Cambrian Burgess Shale-type faunas of North America. *Palaeontology* **36**, 593–635.

Crimes, T. P., Insole, A. and Williams, B. J. P. 1995. A rigid bodied Ediacaran biota from Upper Cambrian strata in Co. Wexford, Eire. *Geological Journal* **30**, 89–109.

Dewel, R. A., Dewel, W. C. and McKinney, F. K. 2001. Diversification of the Metazoa: ediacarans, colonies, and the origin of eumetazoan complexity by nested modularity. *Historical Biology* **15**, 193–218.

Fedonkin, M. A. 1990. Precambrian metazoans. 17–24. In: *Palaeobiology: a Synthesis.* DEG Briggs, PRCrowther (eds.). Blackwell Scientific Publications, Oxford.

Ford, T. D. 1958. Pre-Cambrian fossils from Charnwood Forest. *Proceedings of the Yorkshire Geological Society* **31**, 211–217.

Gehling, J. G., Narbonne, G. M. and Anderson, M. M. 2000. The first named Ediacaran body fossil, *Aspidella terranovica. Palaeontology* **43**, 427–456.

Gehling, J. G. and Narbonne, G. M. 2007. Spindle-shaped Ediacara fossils from the Mistaken Point assemblage, Avalon Zone, Newfoundland. *Canadian Journal of Earth Sciences* **44**, 367–387.

Glaessner, M. F. 1961. Pre-Cambrian animals. *Scientific American* **204**, 72–78.

Glaessner, M. F. and Wade, M. 1966. The Late Precambrian fossils from Ediacara, South Australia. *Palaeontology* **9**, 599–628.

Jenkins, R. J. F. and Gehling, J. G. 1978. A review of the frond-like fossils of the Ediacara assemblage. *Records of the South Australian Museum* **17**, 347–359.

Knoll, A. H. 2003. *Life on a Young Planet.* Princeton University Press, Princeton.

McMenamin, A. S. 1998. *The Garden of Ediacara.* Columbia University Press, New York.

Misra, S. B. 1969. Late Precambrian(?) fossils from southeastern Newfoundland. *Geological Society of America Bulletin* **80**, 2133–2140.

Narbonne, G. M. 2004. Modular construction of early Ediacaran complex life forms. *Science* **305**, 1141–1144.

Narbonne, G. M., Dalrymple, R. W. and Gehling, J. G. 2001. *Neoproterozoic Fossils and Environments of the Avalon Peninsula, Newfoundland.* Geological Association of Canada/Mineralogical Association of Canada Joint Annual Meeting, St John's, Newfoundland Fieldtrip B5 Guidebook.

Narbonne, G. M. and Gehling, J. G. 2003. Life after snowball: the oldest complex Ediacaran fossils. *Geology* **31**, 27–30.

Retallack, G. J. 1994. Were the Ediacaran fossils lichens? *Paleobiology* **20**, 523–544.
Seilacher, A. 1989. Vendozoa: organismic construction in the Proterozoic biosphere. *Lethaia* **22**, 229–239.
Selden, P. and Nudds, J. 2004. *Evolution of Fossil Ecosystems.* Manson, London.
Sprigg, R. C. 1947. Early Cambrian (?) jellyfishes from the Flinders Ranges, South Australia. *Transactions of the Royal Society of South Australia* **71**, 212–224.
Sprigg, R. C. 1949. Early Cambrian 'jellyfishes' of Ediacara, South Australia and Mount John, Kimberley District, Western Australia. *Transactions of the Royal Society of South Australia* **73**, 72–99.
Wood, D. A., Dalrymple, R. W., Narbonne, G. M., Gehling, J. G. and Clapham, M. E. 2003. Paleoenvironmental analysis of the late Neoproterozoic Mistaken Point and Trepassey formations, southeastern Newfoundland. *Canadian Journal of Earth Sciences* **40**, 1375–1391.

The Burgess Shale

Chapter 3

Background: The Cambrian Explosion

Although multicellular animals had only appeared at the very end of the Precambrian (Chapter 2), their evolution at the beginning of the Cambrian was so rapid that this event is known as the 'Cambrian Explosion'. In an astonishing orgy of evolution, a period of little more than 10 million years at the beginning of the Cambrian saw the appearance of almost every animal phylum and body plan known today, along with some other bizarre forms which soon became extinct, suggesting that this was an experimental phase of evolution. Approximately 35 different animal phyla are known today; in the Cambrian seas there were undoubtedly several more and some authorities would claim up to 100.

The sudden appearance of this diverse fauna within the geological record has long posed perplexing questions. Darwin supposed that these phyla had been gradually evolving throughout the Precambrian, but had simply not been preserved as they were entirely soft-bodied. Perhaps they simultaneously acquired preservable hard shells or skeletons at the onset of the Cambrian (in response to a critical change in atmospheric oxygen or oceanic chemistry) so that the Cambrian Explosion was no more than an artefact of preservation?

However, the discovery of the Precambrian Ediacaran biota (Chapter 2) in the latter half of the twentieth century showed that while soft-bodied animals did exist at the very end of the Precambrian, these were mostly primitive annelids and cnidarians and were unlikely ancestors for the characteristic Cambrian animals such as archaeocyathids, brachiopods, trilobites, and mollusks. Moreover, even soft-bodied animals should leave trace fossils, which are also absent from all but the latest Precambrian sediments.

The first wave of the evolutionary explosion is evident in the basal Cambrian Tommotian Stage with the sudden appearance of many small shelly fossils. It may be, as discussed by Fortey (1997) and Conway Morris (1998), that these did have soft-bodied ancestors with a long Precambrian history, but which were so small that neither their bodies nor their traces would be preserved. Even so it is unlikely that such tiny animals, often only a few millimetres in length, were equipped with the huge range of body plans that were evident later in the Cambrian; perhaps these 'small shellies', as they are known, are simply the hard spikes or spines of soft-skinned animals.

Very soon after their appearance the main burst of evolution began and a variety of possible triggers has been put forward. Gradually increasing oxygen levels after 3 billion years of photosynthesis by cyanobacteria and later by plants may have allowed the evolution of more mobile and larger, complex animals which were able to exploit empty niches significantly devoid of competitors. Alternatively, the emergence of the first predators may have initiated the first 'arms race', in which animals had either to escape by moving faster and/or getting bigger or to defend themselves by developing hard shells. Changes in continental configuration (and therefore of ocean currents) have also been suggested. Or maybe genetic mechanisms were simply more flexible at that time, leading to accelerated diversification.

Much of our knowledge of the fauna and flora during the Cambrian Explosion (called 'the crucible of creation' by Conway Morris, 1998) is gleaned from perhaps the best known of all Fossil-Lagerstätten – the Burgess Shale of British Columbia in Canada. By an accident of geological history, this thin layer of shale has provided a window into the richness of a Middle Cambrian marine ecosystem at a most vital time in the evolution of life on Earth. Here, by a process still not completely understood, decay was arrested so that the complete diversity of the Cambrian seas, including many soft-bodied animals (with their internal organs and muscles) has been exquisitely preserved.

These are the 'weird wonders' popularized by Stephen Jay Gould (1989) in his book *Wonderful Life*. And if one considers that approximately 85% of Burgess Shale genera are entirely soft-bodied and thus absent from other Cambrian assemblages, it becomes apparent just how misleading the fossil record would be had this particular Lagerstätte not been preserved or discovered.

History of discovery of the Burgess Shale

It was an American, Charles Doolittle Walcott, then Secretary of the Smithsonian Institution in Washington DC, who first discovered the Burgess Shale, high in the Canadian Rockies (**28**). The romantic story suggests that at the end of the 1909 field season his wife's horse, descending the steep Packhorse Trail that leads off the ridge between Mount Wapta and Mount Field, in what is now the Yoho National Park, stumbled on a boulder. Walcott dismounted to clear the track and on splitting the offending boulder revealed a fine specimen of the soft-bodied 'lace crab', *Marrella*, glistening as a silvery film on the black shale.

Unfortunately his diary does not support this story, but certainly Walcott discovered the first fossils in September 1909 and began excavating in earnest during the summer of 1910. Annual field seasons with his family continued until 1913 and he returned in 1917, 1919, and 1924. By the time of his death in 1927 he had amassed a collection of 65,000 specimens, which were transported back to Washington DC, where they remain today in the Smithsonian Institution.

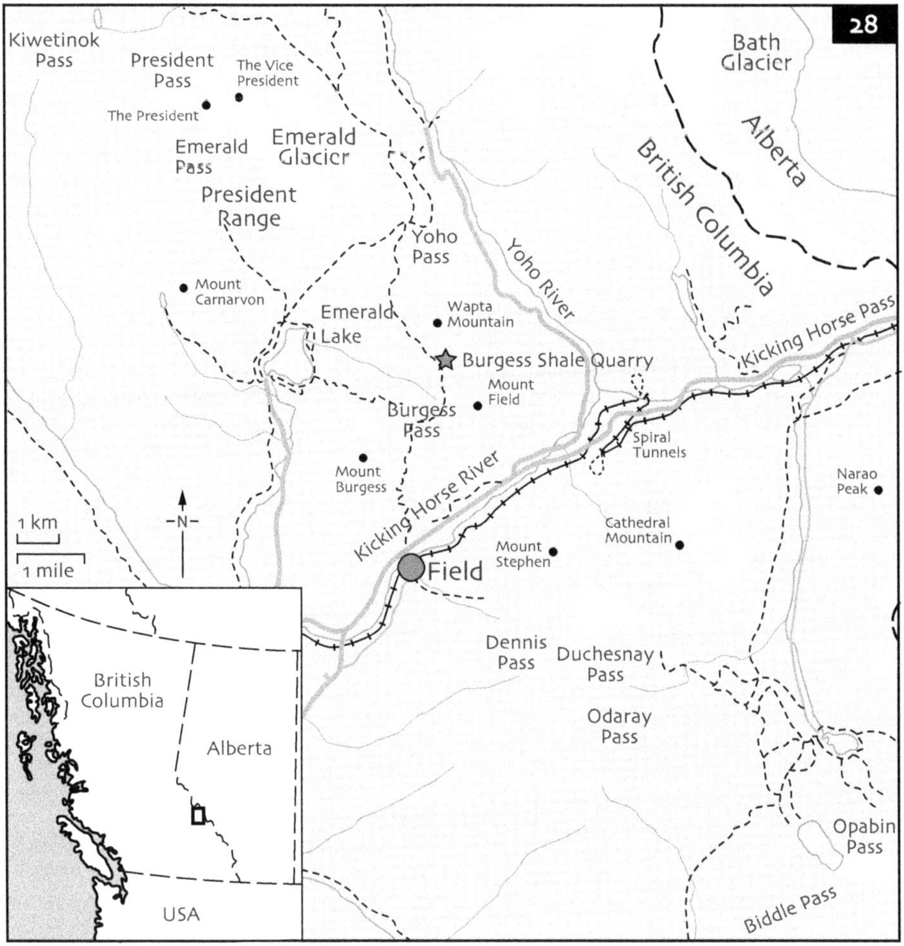

28 Locality map showing the position of the Burgess Shale Quarry within Yoho National Park in the Canadian Rockies.

By 1912 Walcott had published many of his finds and, of the 170 species currently recognized from the Burgess Shale, over 100 were described by Walcott himself.

Walcott's Burgess Shale Quarry (**29, 30**) lies at 2,300 m (7,500 ft), just below the top of Fossil Ridge, which connects Mount Wapta and Mount Field. It is a stunning location. Looking west from the quarry, which remains snow-filled even in the summer, a breath-taking panorama of snow-capped mountains, glaciers, lakes, and forests is revealed. Mount Burgess points skywards, the vivid green Emerald Lake is to its right, with the Emerald Glacier poised above it. The site of Walcott's camp can be made out on the High Line Trail below.

Walcott also collected higher up the mountain at a site now called the Raymond Quarry after its detailed excavation in the 1930s by Professor Percy E Raymond of Harvard University, whose collection now resides at the Museum of Comparative Zoology at Harvard. Apart from the work of Alberto Simonetta, an Italian biologist, who made detailed descriptions of some of the Burgess trilobites in the 1960s, little further work on these remarkable fossils was carried out until the notion of a complete restudy of the Burgess fauna was proposed by Professor Harry Whittington in 1966.

Whittington is an Englishman, but in 1966 was Professor of Paleontology at Harvard University. He managed to persuade the Geological Survey of Canada that a restudy was timely and collecting trips in 1966 and 1967 produced over 10,000 new specimens, although few new taxa. In 1966 Whittington moved from Cambridge, Massachusetts, to Cambridge, England, and took the project with him. Very soon he recruited two young postgraduates to assist in the huge task of redescribing the Burgess animals, Derek Briggs, taking on the arthropods, and Simon Conway Morris, the worms.

The Cambridge team made painstaking dissections, drawings, and detailed photographs of the Burgess fossils and revealed a wealth of new data unseen by previous workers. Whittington produced a detailed restudy of Walcott's first find, the 'lace crab', *Marrella*, the most common animal from the Burgess fauna, and set new standards for the description of Burgess fossils. Briggs and Conway Morris worked respectively on two of the real enigmas of the Burgess Problematica, *Anomalocaris* and *Canadia sparsa* (later renamed *Hallucigenia*) and by remarkable detective work were able to elucidate the true affinities of these most bizarre Burgess animals.

In 1975 Desmond Collins of the Royal Ontario Museum (ROM) in Toronto obtained permission from the Park Rangers to collect loose material from the ridge. This party returned in 1981 and 1982, this time to look for new localities, and demonstrated that the fossiliferous beds were actually more extensive than had been thought, with 10 or more new locations, both above and below Walcott's quarry, all along the line of the Cathedral Escarpment.

In 1981 the site was designated a World Heritage Site by UNESCO. The work of the ROM team continued throughout the 1980s and 1990s with some remarkable finds.

Stratigraphic Setting and Taphonomy of the Burgess Shale

The Burgess Shale Formation is about 270 m (886 ft) thick and includes 10 members (Fletcher and Collins, 1998), one of which, the Walcott Quarry Shale Member, includes the productive layers in Walcott's quarry which have yielded the well-preserved, soft-bodied fossils. This thick formation is the deep-water lateral equivalent of the much thinner platform facies of the Stephen Formation (**31**) and is Middle Cambrian in age, approximately 510 million years old.

Just to the north of Walcott's quarry the dark shales of the Burgess Shale Formation abruptly disappear and abut against much lighter-coloured dolomites belonging to the Cathedral Formation with an almost vertical contact (**31**).

29 Walcott's Burgess Shale Quarry at 2,300 m (7,500 ft) in Yoho National Park, Canada; Mount Wapta can be seen in the background.

30 Detail of laminated shale in the Burgess Shale Quarry, showing fining-upwards sequence with coarse, orange layers at the base and finer, grey layers above. (Measurements indicate centimetres below Walcott's 'Phyllopod Bed'.)

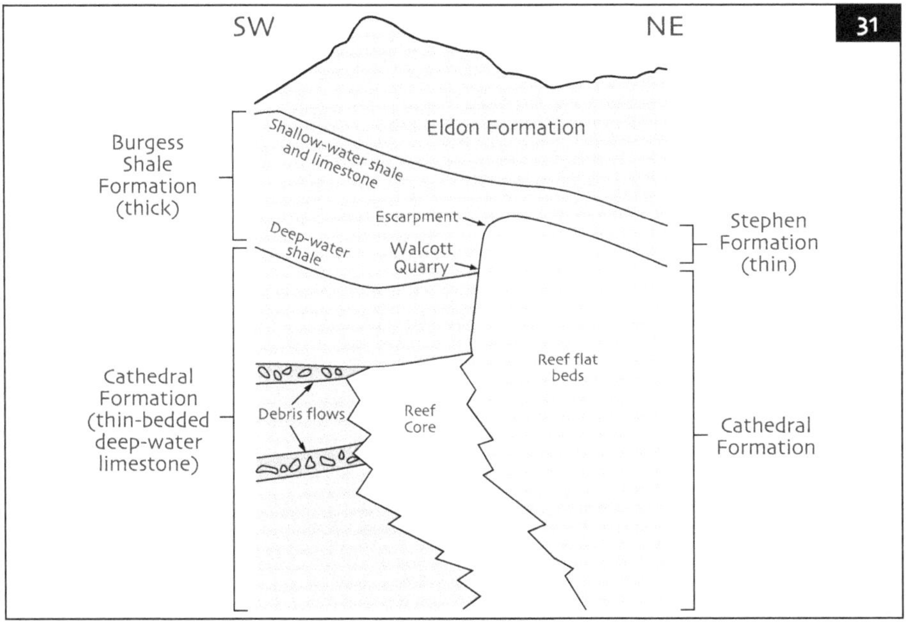

31 Diagram to show the stratigraphy and structure of the Middle Cambrian succession on Mount Field (after Briggs *et al.*, 1994, Fletcher and Collins, 1998).

The fact that the overlying Eldon Formation passes uninterrupted across this vertical contact shows that it is not a tectonic fault, but instead that it was an original feature of the Cambrian seabed, i.e., a near-vertical submarine cliff. It is significant that the Burgess Shale fossils are always found at the foot of this submarine cliff.

The conventional interpretation is that this cliff represents the margins of an algal reef, the top of which formed a shallow carbonate platform with well-lit waters free of terrigenous sediment. The Burgess animals lived on or in the mud in the deeper, darker water at the foot of the cliff. The seabed sloped away from the cliff into still deeper water which was anoxic and hostile to life.

Various pieces of evidence suggest that the Burgess animals were not preserved in the area in which they were living. The first is the lateral continuity over large distances of the thin beds of shale, with no evidence of bioturbation of the sediment by crawling or burrowing organisms. The second is the fact that the Burgess fossils are found lying at all possible angles within the shale, some even head-first into the sediment. Finally, examination of the thin shale beds in Walcott's Quarry reveals that each shows a definite fining-upwards sequence with coarse, orange layers at the base and finer, dark grey layers above, such that each bed represents a separate event, i.e. a separate influx of sediment (**30**).

The accepted theory of the deposition of the Burgess animals is that from time to time storms, earth movements, or simply instability of the wet sediment pile sent the mud at the foot of the cliff down into the hostile basin in a rapid cloud of sediment, carrying with it the unsuspecting animals which had no time to escape. Conway Morris (1986, Figure 1) illustrated a 'pre-slide' and 'post-slide' environment, the former representing the area in which the animals were living, and the latter the area to which they were transported and preserved (now represented by Walcott's Quarry). Both appear to have been close to the foot of the cliff, suggesting that the turbidity currents flowed downslope parallel to the cliff. The slide was followed by quiet conditions allowing the fine sediment to settle, giving rise to the fine layering of the shale seen today.

How the Burgess Shale animals have been preserved is still not entirely known. Two of the common prerequisites for soft-tissue preservation were undoubtedly partly responsible, namely rapid, catastrophic burial in a fine sediment, and deposition on a sea floor deficient in oxygen. Such a toxic 'post-slide' environment would have excluded scavengers, and when the cloud of sediment settled the carcasses would have been completely entombed with their body cavities infilled by mud. However, anaerobic microbes can break down soft muscle tissue relatively quickly even in the absence of oxygen, and some other factor must have prevented such microbial action.

Butterfield (1995) isolated tissues of various soft-bodied Burgess animals and showed that in many cases they were composed of altered original organic carbon. This was coated, however, by a thin film of calcium aluminosilicate, similar to mica, explaining the silvery appearance of the Burgess fauna under incident light. He suggested that this coating originated from the clay minerals in the mud in which the animals were buried, and that these minerals inhibited bacterial decay, perhaps by preventing reactions of enzymes. This type of preservation is exceptional; normally very soft tissue such as muscle and intestine can only be preserved if it is replaced by another mineral during early diagenesis.

Description of the Burgess Shale biota

Vauxia (Phylum Porifera)

Vauxia is a bush-like, branching sponge (**32**), which does not have discrete spicules, but is composed of a tough spongin-like framework, explaining why it is the most common of the Burgess sponges.

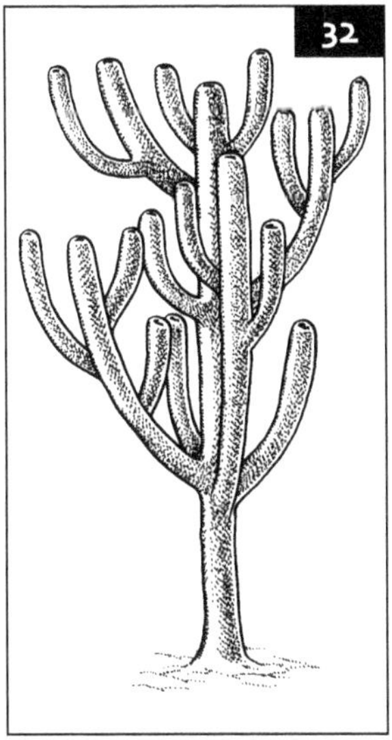

32 Reconstruction of *Vauxia*.

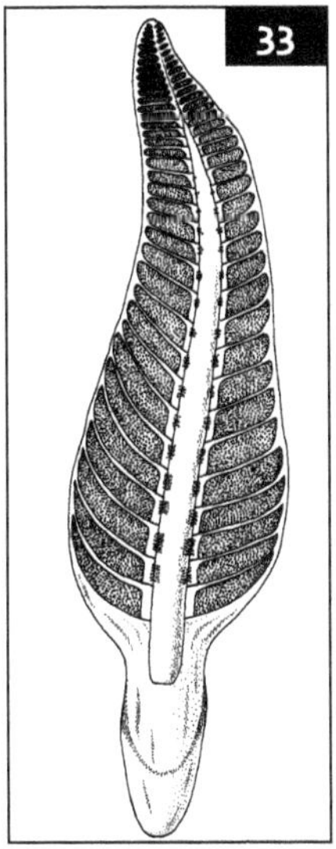

33 Reconstruction of *Thaumaptilon*.

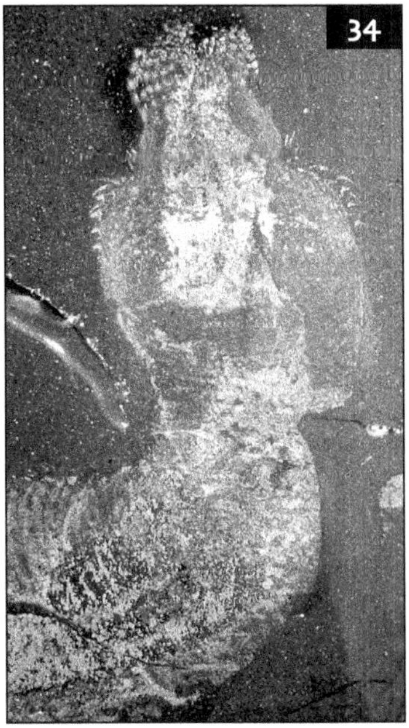

34 The priapulid worm *Ottoia prolifica*, showing anterior proboscis with spines (USNM). Diameter <10 mm (0.4 in).

Thaumaptilon (Phylum Cnidaria; Order Pennatulacea)

A rare constituent of the sessile Burgess fauna, this possible sea pen (**33**) is nonetheless important as it is a survivor of the Precambrian Ediacara fauna (Chapter 2), resembling the genus *Charnia* (see **22**). It has a broad central axis with up to 40 branches, each housing hundreds of individual star-like polyps (see Conway Morris, 1993).

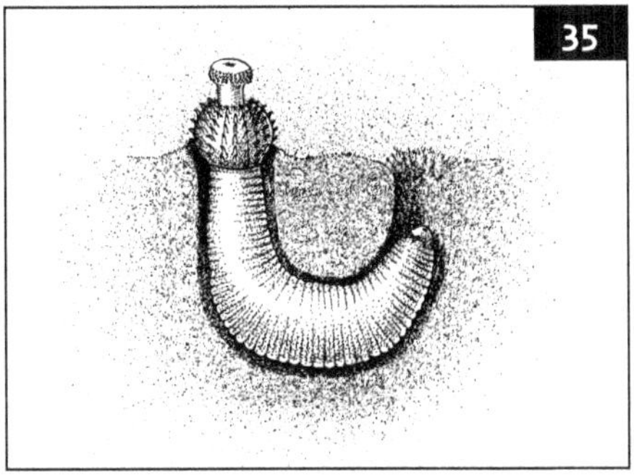

35 Reconstruction of *Ottoia*.

Ottoia (Phylum Priapulida)

This is the most abundant of the mud-dwelling priapulid worms (**34, 35**), especially in the higher beds of the Burgess Shale. Priapulids are carnivorous animals, rare today, but common in the Cambrian seas. They had a bulbous anterior proboscis surrounded by vicious hooks and spines. At the end of the proboscis is a mouth with sharp teeth; their stomachs often reveal the last meal, which may includes hyoliths, brachiopods, and even other specimens of *Ottoia*, for these animals were cannibals. They are commonly fossilized in a U-shape, suggesting that they lived in U-shaped burrows. However, recently discovered straight specimens may suggest that the U-shape was due to post-mortem contraction (see Conway Morris, 1977a).

Burgessochaeta and *Canadia* (Phylum Annelida; Class Polychaeta)

These are polychaete worms, or bristle worms (**36**), segmented animals with paired appendages each bearing numerous bristles or setae (see Conway Morris, 1979).

Hallucigenia (Phylum Onychophora)

The most celebrated Burgess animal and the classic 'weird wonder' of Stephen Jay Gould (1989) was placed in this new genus by Conway Morris to reflect its dream-like quality as it seemed unlike any other

36 The polychaete worm *Canadia spinosa* (USNM). Length about 30 mm (1.2 in).

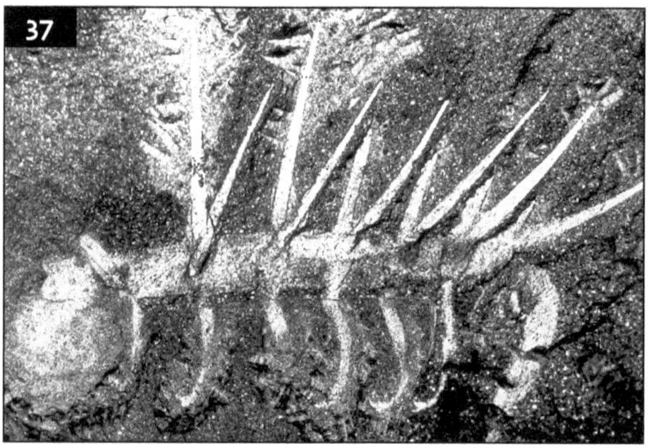

37 The velvet worm *Hallucigenia sparsa* (USNM). Length about 20 mm (0.8 in).

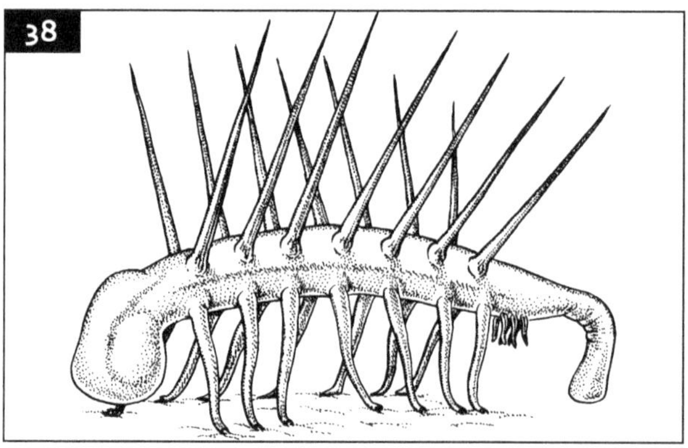

38 Reconstruction of *Hallucigenia*.

known animal (**37**, **38**). This was partly because it had been reconstructed upside-down, standing on rigid spines and waving its tentacles in the water. Additional specimens suggested reversing this interpretation and suddenly it was apparent that this genus was a marine velvet worm, a caterpillar-like group called the lobopodians ('lobed feet'). *Hallucigenia* crawled on its fleshy limbs and used its spines for protection as it scavenged for decaying food. *Aysheaia* is another such Burgess lobopod (see Conway Morris, 1977b).

Marrella (Phylum Arthropoda)

A small, feathery arthropod, often known as the 'lace-crab', it is the most common Burgess animal, with over 15,000 specimens discovered, yet it is known from no other Cambrian deposit (**39, 40**). It was the first to be discovered by Walcott and the first to be redescribed by Whittington. A head shield has two pairs of curving spines, while the head has two pairs of antennae. The 20 body segments each bear a pair of identical legs, suggesting that it is a primitive arthropod and could be ancestral to the three major groups of aquatic arthropods (crustaceans, trilobites, chelicerates; see Whittington, 1971).

39 The arthropod *Marrella splendens* (MM). Length 20 mm (0.8 in).

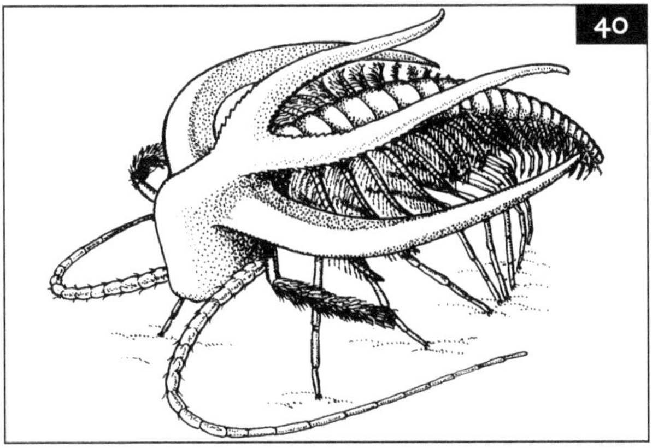

40 Reconstruction of *Marrella*.

Sanctacaris (Phylum Arthropoda, Class Chelicerata)

This is the most important specimen to be discovered by Collins and the ROM team as it represents the earliest known example of a chelicerate, the group containing the spiders and scorpions (**41**). The large head shield protecting six head appendages, five of which were spiny claws to assist in capturing prey, gave it the nickname of 'Santa Claws'! (see Briggs and Collins, 1988).

Anomalocaris (Problematica or Phylum Arthropoda?)

The monster predator of the Burgess Shale, *Anomalocaris,* does not resemble any known animal and has long been considered as an example of a short-lived experimental arthropod-like phylum (**42**). The prey-grasping anterior appendages were originally thought to represent the segmented abdomen of a crustacean (hence the generic name meaning 'strange crab'), then were interpreted as a set of paired limbs of a giant arthropod, while the circular mouth parts were originally interpreted as the jellyfish *Peytoia.* More complete specimens revealed this to be the largest known of the Burgess animals, reaching lengths of up to 1 m (3.3 ft). The head has a pair of large eyes, the trunk is covered with flap-like structures, and the tail is a spectacular fan (see Whittington and Briggs, 1985).

Opabinia (Problematica)

This truly strange animal had five eyes on the top of its head and a long, flexible proboscis which terminated in a number of spines, apparently a grasping organ (**43, 44**). Each of the body segments possesses lateral lobes with gills and a strange tail was formed by three posterior flaps. Some authorities now believe that it may be related to *Anomalocaris* and that both may be arthropods (see Whittington, 1975).

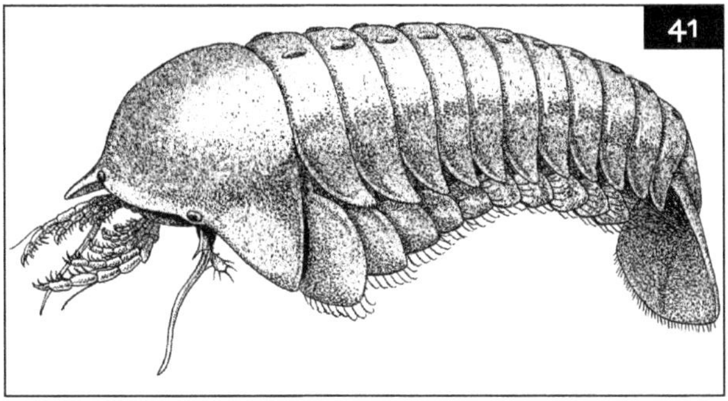

41 Reconstruction of *Sanctacaris*.

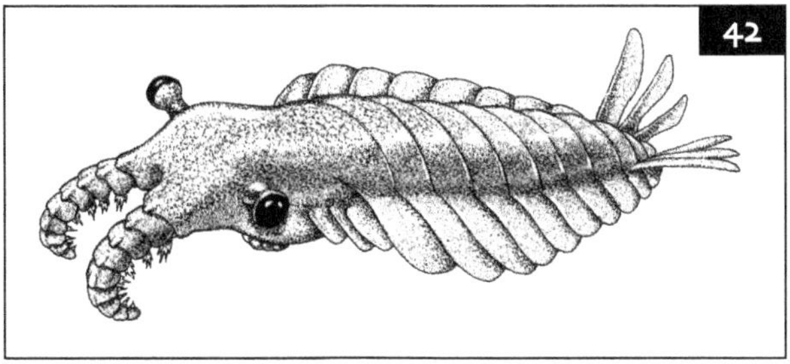

42 Reconstruction of *Anomalocaris*.

43 The problematical *Opabinia regalis* (USNM). Length about 65 mm (2.5 in).

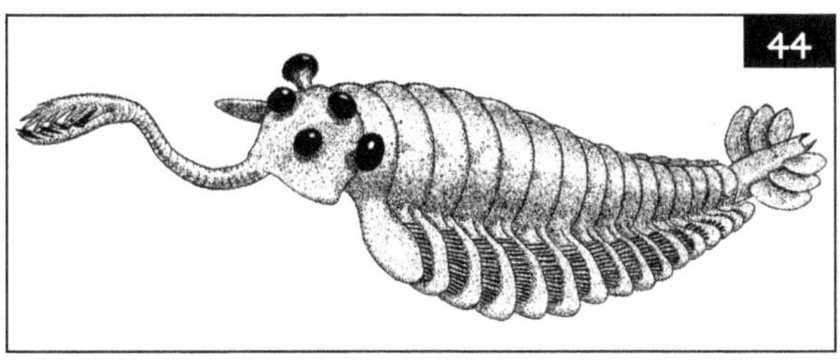

44 Reconstruction of *Opabinia*.

Wiwaxia (Problematica or Phyllum Mollusca)

This strange mud-crawling animal was protected from predators by its dorsal surface being covered by a coat of scale-like sclerites and a double row of pointed spines (**45**). The ventral surface was a soft foot similar to that of a slug or snail and from the open mouth protruded a radula, also reminiscent of mollusks; indeed Caron *et al* (2006) interpreted this genus as a stem group mollusk. The microstructure of the sclerites, however, is more akin to that of polychaete annelids. The true affinities of *Wiwaxia* thus remain in doubt, although it is almost certainly related to the halkieriids (**46**) (see section on Sirius Passet below and Conway Morris, 1985).

Pikaia (Phylum Chordata)

An inconspicuous but vital element of the Burgess fauna, *Pikaia* possessed a stiff rod along its dorsal margin (**47**), which suggests that it is a primitive chordate, the phylum to which Man and all vertebrates belong, and showing that even our

45 The problematical *Wiwaxia corrugata* (USNM). Length about 35 mm (1.4 in).

46 The halkieriid *Halkieria evangelista*, from Greenland (GMC). Length 60 mm (2.4 in).

ancestors were present during the Cambrian Explosion. A narrow anterior end has a pair of tentacles, while the posterior is expanded into a fin-like tail. The cone-in-cone arrangement of the muscles is often clearly preserved.

Paleoecology of the Burgess Shale

The Burgess Shale represents a marine, benthic community living in, on, or just above the muddy seabed at the foot of a submarine cliff, where the mud was banked sufficiently high to be clear of stagnant bottom waters. The marine basin faced the open sea and was situated within the tropical zone at about 15°N. The presence of photosynthesizing algae suggests that the depth was not much more than 100 m (330 ft).

Paleoecological analysis by Conway Morris (1986) examined over 30,000 slabs of shale with 65,000 fossils. Approximately 10% of the biota consisted of benthic infauna, i.e., living in the sediment itself, this community being dominated by the burrowing priapulid worms such as *Ottoia*, *Selkirkia*, and *Louisella*, and polychaete worms such as *Burgessochaeta* and *Canadia*.

The vast majority of Burgess animals (c. 75%) consisted of benthic epifauna, living on the sediment surface, and divided into the fixed or sessile epifauna (c. 30%) and the vagrant epifauna (c. 40%), that walked or crawled across the seabed. The sessile biota consisted mainly of sponges, such as the branching *Vauxia*, and others, such as *Choia* and *Pirania* adorned with sharp, glassy spicules which both supported the skeleton and protected against predators. The sea pen, *Thaumaptilon*, was a rare member of the sessile epifauna, as was the enigmatic animal *Dinomischus*, projecting just 1 cm (0.4 in) above the mud on a thin stalk and looking like a small flower.

The vagrant epifauna was more varied, but was dominated by arthropods, only a small proportion of which were trilobites, such as *Olenoides* and the soft-bodied *Naraoia*. It also included the ubiquitous 'lace crab', *Marrella*, and a number of small, scavenging lobopods, most notably *Hallucigenia* and *Aysheaia*. The most bizarre mud crawler was *Wiwaxia*, which moved across the sediment using its slug-like foot.

Animals which lived above the mud surface were fewer in number, comprising only about 10% of the Burgess biota, simply because they were more able to escape the mud flow by swimming away. The nektobenthic animals (near-bottom swimmers) were, however, within the range of turbidity flows and included the tiny chordate *Pikaia*. The medusoid *Eldonia*, which is more probably related to the holothurians (sea cucumbers) than to true jellyfish, may be regarded as nektobenthos and/or a planktonic floater. Nekton (forms swimming above the substrate) include the giant predator *Anomalocaris* and the enigmatic

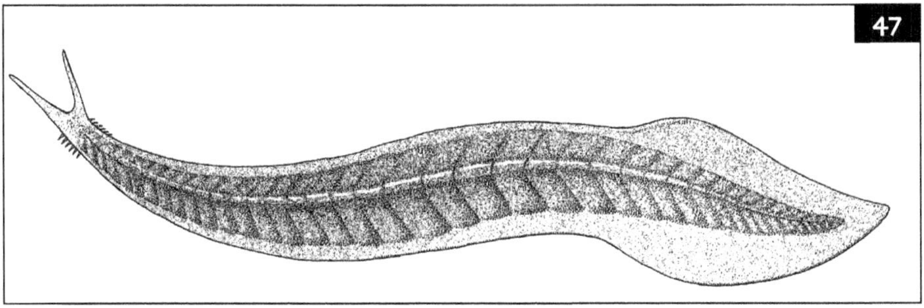

47 Reconstruction of *Pikaia*.

Opabinia, which used its nozzle-like proboscis to search for food. Trophic analysis by Conway Morris (1986) identified several feeding types including: filter feeders, dominated by the sponges; deposit feeders, dominated by arthropods (deposit collectors) and mollusks (deposit swallowers); scavengers such as the lobopods *Hallucigenia* and *Aysheaia*; and lastly, predators such as the giant *Anomalocaris,* large arthropods (such as *Sidneyia*) and the priapulids. The important role of predation evident in the Burgess Shale ecological framework provides the main contrast with other Cambrian (shelly) assemblages.

COMPARISON OF THE BURGESS SHALE WITH OTHER CAMBRIAN BIOTAS

Burgess Shale-type assemblages have since been discovered at about 40 localities worldwide, but two are of special significance.

Sirius Passet, northern Greenland

Discovered in 1984 by the Geological Survey of Greenland near J P Koch Fjord in Peary Land, north Greenland, this locality, now known as Sirius Passet, was first seriously collected by a 1989 expedition. It soon became apparent that this was another predominantly soft-bodied Cambrian fauna, also seemingly deposited in deep water muds adjacent to a shallow carbonate bank, and more than 3,000 specimens were collected from what is now termed the Buen Formation. Significantly, however, this fauna is of Lower Cambrian age (Atdabanian), somewhat older than the Burgess Shale, revealing that the Cambrian Explosion was well underway by this time.

One of the first and most intriguing specimens to be discovered at Sirius Passet was *Halkieria,* a slug-like animal (**46**) with a dorsal coat of scale-like sclerites just as in *Wiwaxia* (**45**). But, at either end of the long body is a shell looking remarkably like an inarticulate brachiopod (see Conway Morris and Peel, 1990; Conway Morris, 1998, Figure 86). Perhaps these animals are telling us that mollusks, annelids *and* brachiopods are phylogenetically closer than had been supposed? Molecular biology supports this conclusion.

Chengjiang, southern China

'Discovered' in the same year, 1984 (although actually known since 1912), the Chengjiang Fossil-Lagerstätte, best exposed at Maotianshan, near Kunming, in Yunnan Province, southern China, also yields abundant soft-bodied 'Burgess Shale-type' fossils (see Chen and Zhou, 1997; Hou *et al.,* 2004). Although of similar age to the Sirius Passet assemblage (Lower Cambrian, Atdabanian Stage), many of the characteristic Burgess animals are known from Chengjiang (including complete specimens of *Hallucigenia* and *Anomalocaris*), in addition to new Chinese genera of arthropods, worms, sponges, brachiopods and so on. The faunal similarity is remarkable because the South China craton would have been thousands of kilometres from the continent of Laurentia (comprising North America and Greenland).

Among several new groups, a new phylum, *Vetulicolia,* was proposed by Shu *et al.* (2001) to include segmented, arthropod-like metazoans which had obvious gill slits, suggesting a deuterostome affinity. But perhaps the most surprising element of the fauna was reported by Shu *et al.* (1999): the discovery of agnathan fish (previously known from the Lower Ordovician), illustrating that even the vertebrates appeared during the Cambrian Explosion.

Fossil preservation is spectacular with reddish-purple impressions on fine orange shale, and again appears to be the result of rapid, live burial in catastrophic turbidity flows. The Chengjiang animals probably lived in the area of their burial, which was adjacent to a delta front. The 50 m (160 ft) thick Maotianshan Shale Member of the Yu'anshan Formation consists of thin, graded mudstone layers, recording short episodic sedimentation events.

Further Reading

Briggs, D. E. G. and Collins, D. 1988. A Middle Cambrian chelicerate from Mount Stephen, British Columbia. *Palaeontology* **31**, 779–798.

Briggs, D. E. G., Erwin, D. H. and Collier, F. J. 1994. *The Fossils of the Burgess Shale.* Smithsonian Institution Press, Washington DC.

Butterfield, N. J. 1995. Secular distribution of Burgess Shale-type preservation. *Lethaia* **28**, 1–13.

Caron, J.-B., Scheltema, A., Schander, C. and Rudkin, D. 2006. A soft-bodied mollusc with radula from the Middle Cambrian Burgess Shale. *Nature* **442**, 159–163.

Chen, J. and Zhou, G. 1997. Biology of the Chengjiang fauna. *Bulletin of the National Museum of Natural Science* **10**, 11–105.

Conway Morris, S. 1977a. Fossil priapulid worms. *Special Papers in Palaeontology* **20**, 1–95.

Conway Morris, S. 1977b. A new metazoan from the Cambrian Burgess Shale of British Columbia. *Palaeontology* **20**, 623–640.

Conway Morris, S. 1979. Middle Cambrian polychaetes from the Burgess Shale of British Columbia. *Philosophical Transactions of the Royal Society of London*, Series B **285**, 227–274.

Conway Morris, S. 1985. The Middle Cambrian metazoan *Wiwaxia corrugata* (Matthew) from the Burgess Shale and *Ogygopsis* Shale, British Columbia. *Philosophical Transactions of the Royal Society of London*, Series B **307**, 507–582.

Conway Morris, S. 1986. The community structure of the Middle Cambrian Phyllopod Bed (Burgess Shale). *Palaeontology* **29**, 423–467.

Conway Morris, S. 1993. Ediacaran-like fossils in Cambrian Burgess Shale-type faunas of North America. *Palaeontology* **36**, 593–635.

Conway Morris, S. 1998. *The Crucible of Creation.* Oxford University Press, Oxford.

Conway Morris, S. and Peel, J. S. 1990. Articulated halkieriids from the Lower Cambrian of north Grenland. *Nature* **345**, 802–805.

Fletcher, T. P. and Collins, D. H. 1998. The Middle Cambrian Burgess Shale and its relationship to the Stephen Formation in the southern Canadian Rocky Mountains. *Canadian Journal of Earth Science* **35**, 413–436.

Fortey, R. 1997. *Life: an Unauthorised Biography.* Flamingo, London.

Gould, S. J. 1989. *Wonderful Life: the Burgess Shale and the Nature of History.* Norton, New York.

Hou, X. G., Aldridge, R. J., Bergström, J., Siveter, D. J., Siveter, D. J. and Feng, X. H. 2004. *The Cambrian Fossils of Chengjiang, China.* Blackwell, Oxford.

Shu, D. G., Luo, H. L., Conway Morris, S. *et al.* 1999. Lower Cambrian vertebrates from south China. *Nature* **402**, 42–46.

Shu, D. G., Conway Morris, S., Han, J. *et al.* 2001. Primitive deuterostomes from the Chengjiang Lagerstätte (Lower Cambrian, China). *Nature* **414**, 419–424.

Whittington, H. B. 1971. Redescription of *Marrella splendens* (Trilobitoidea) from the Burgess Shale, Middle Cambrian, British Columbia. *Bulletin of the Geological Survey of Canada* **209**, 1–24.

Whittington, H. B. 1975. The enigmatic animal *Opabinia regalis*, Middle Cambrian, Burgess Shale, British Columbia. *Philosophical Transactions of the Royal Society of London*, Series B **271**, 1–43.

Whittington, H. B. and Briggs, D. E. G. 1985. The largest Cambrian animal, *Anomalocaris*, Burgess Shale, British Columbia. *Philosophical Transactions of the Royal Society of London*, Series B **309**, 569–609.

Beecher's Trilobite Bed

Chapter 4

Background

Between the extraordinary biotas of the Early – Middle Cambrian (such as Burgess Shale, Chengjiang, Chapter 3) and the beautifully preserved arthropods of the late Silurian Bertie Waterlimes described in Chapter 5, the fossil record shows few major Lagerstätten. The Ordovician Period is particularly barren of exceptional biotas. However, one Ordovician horizon, Beecher's Trilobite Bed, part of the Upper Ordovician Utica Basin shales of upper New York State (**48**), has long been famous for its trilobites with appendages exquisitely preserved in iron pyrite (FeS_2). While there are many occurences of fossils preserved in pyrite in the fossil record, e.g. beautiful golden goniatite shells in Coal Measure marine bands of northern England, pyrite preservation of soft-parts anatomy is extremely rare. Another example of soft-part preservation in pyrite is in the Devonian Hunsrück Slate of Germany (Bartels *et al.*, 1998; Selden and Nudds, 2004, Chapter 4).

As we saw in the last chapter, life in Cambrian seas was remarkably different from today. While probably most of the animals in the Burgess Shale belong to phyla which are still in existence, many species did not extend past the end of the Cambrian. The relative abundances of phyla also differed greatly from a typical marine scene today. The Cambrian sea was rich in siliceous sponges (Porifera) and inarticulate brachiopods with phosphatic shells; priapulid worms dominated the soft-bodied fauna – in contrast, the polychaete annelids are the most familiar worms today.

Jack Sepkoski of the University of Chicago was the first scientist to identify a distinct Cambrian Fauna (Sepkoski, 1979). He noticed how after the end of the Cambrian Period, many of the typical animals of the Cambrian had either became extinct or (mostly) persisted at much lower diversity than before. Their ecological niches had been taken over by new forms. For example, inarticulate brachiopods still persisted, even to this day, but articulate brachiopods became the dominant sea-floor shells; the sessile fauna had been dominated by sponges, but in the Ordovician Period corals and echinoderms were outcompeting them.

Our main interest in this chapter is with the trilobite arthropods. Two kinds of trilobites were typical of Cambrian seas: large, multisegmented redlichiids, and tiny, two- or three-segmented agnostids. The former lay flat on the sea bed while the latter most likely belonged to the plankton (floating forms). Other trilobite orders emerged during the Cambrian, but these two were present from the beginning. Redlichiids did not extend beyond the Cambrian, but the agnostids continued to the end of the Ordovician

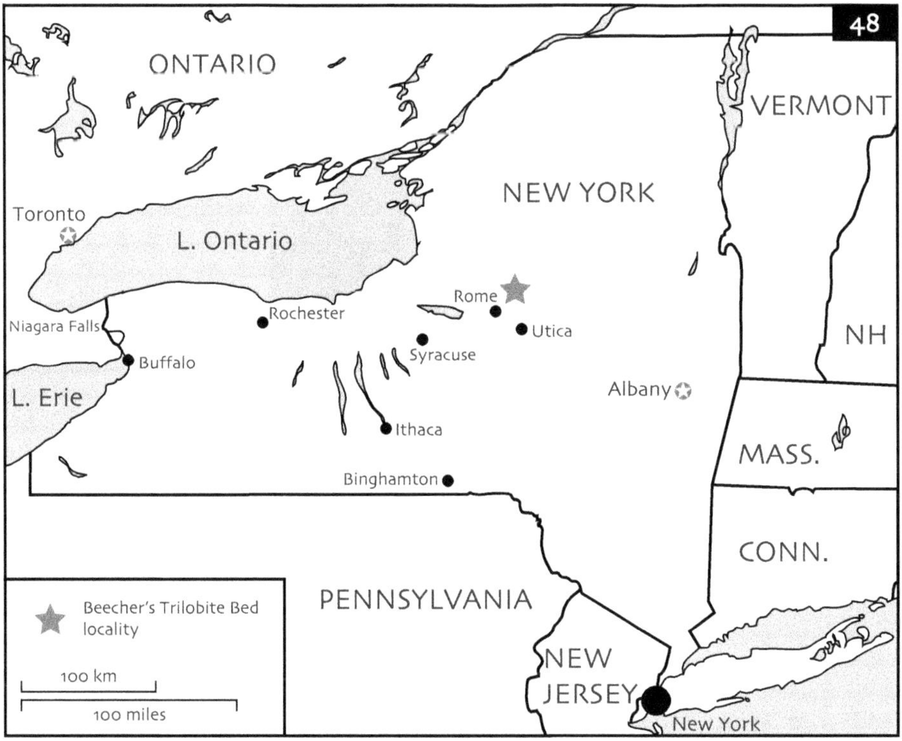

48 Map of New York showing the position of Beecher's Trilobite Bed near Rome (after Bottjer *et al.*, 2001).

Period. So, at the start of the Ordovician, there were new groups of trilobites, such as corynexochids, asaphids, and ptychopariids, which had fewer, determinate numbers of segments and inhabited a wider variety of ecological niches. They belonged to Sepkoski's Paleozoic Fauna.

History of discovery of Beecher's Trilobite Bed

Beecher's Trilobite Bed is something of a misnomer: it was not discovered by Charles Beecher, and although its fame is entirely due to the trilobites it contains, the landmark publication on the anatomy of these animals was published not by Beecher but by Percy Raymond (Raymond, 1920). Nevertheless, the story of the discovery of the bed, and how it got its name and fame, is fascinating.

Like all arthropods, trilobites have an external cuticle made of an organic complex of chitin (a polysaccharide) and arthropodin (a protein). In addition, on the dorsal side of the body the cuticle is impregnated with hard calcium carbonate, like the shell of a crab or lobster. Because the legs and other ventral parts of the animal's anatomy are

49 An old photograph of Charles Beecher (right) at his excavation, together with his brother Coleman Beecher (centre), his uncle Henry Downer (left) and, collecting fossils behind, 'Doc' Randall (YPM).

not calcified, when the animal dies or molts its skin, the hard dorsal surface stands a much greater chance of being fossilized than the soft parts. Nearly all fossil trilobites show the hard dorsal surface only, and trilobite soft parts are very rarely preserved. William S Valiant, a keen amateur fossil collector from Rome, New York, was aware of a discovery by Charles Walcott (who later found the Burgess Shale, Chapter 3) of trilobites with appendages in thin sections of the Trenton Limestone of New York (Walcott, 1876, 1879). In his searches for fossils along Six Mile Creek at Cleveland's Glen near Rome, New York, Valiant had found a chip of shale with what looked tantalizingly like a trilobite appendage on it. Curious as to where this piece of shale originated, and hoping to find some more trilobites with appendages, in 1884 he started a concerted search along the creek. At that time, trilobite legs were known, but scientists were perplexed that none had shown any sign of antennae which, by comparison with crustaceans, it was widely believed trilobites should possess. Valiant hoped to be the first to find trilobites with antennae and, after searching the creek for 8 years, in 1892 he and his half-brother Sid Mitchell located the bed which yielded the pyritized fossils and there were trilobites with antennae!

Valiant wrote to several New York paleontologists about his find, including the New York State Geologist, James Hall, but received no replies. Eventually, through the help of a Rome businessman Kingsley and Professor James F Kemp at Columbia College, some specimens reached Kemp's student WD Matthew who wrote the first description of trilobites with antennae (Matthew, 1893). Some more specimens reached Professor Marsh at Yale University (see Chapter 9), who passed them to his young colleague Charles Beecher. Walcott visited the site and published a note on the trilobite appendages from there (Walcott, 1894). So how did the site become known as Beecher's Bed? In 1893, Beecher took out a lease on the land with the exclusive rights to dig for fossils. In 1895, Walcott searched the adjacent plot upstream of

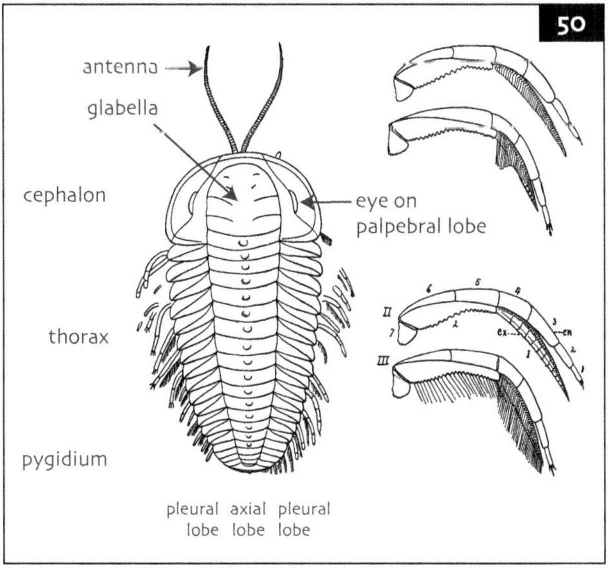

50 Beecher's (1893) drawing of *Triarthrus,* annotated to show basic trilobite morphology. Whole animal is about 3 cm (1.2 in) long (excluding antennae). Diagrammatic appendages shown at right are from thoracic segments 2 and 3 from another specimen; note the two branches: upper, leg branch (endopodite) and lower, gill branch (exopodite). The appendage labelled II has the setae removed from the gill branch; that labelled III shows all setae. Legs are shown at a magnification of 4× the size of the whole trilobite.

Beecher's quarry. Walcott removed 365 cubic yards of shale in 6 days, and took the trilobites back to Washington DC (Whiteley, 1998); meanwhile, Beecher had been collecting material for the Yale Peabody Museum (**49**).

Charles Emerson Beecher was the first invertebrate paleontologist at Yale and was appointed Curator and Professor of Geology in 1891. He developed one of the earliest classifications for trilobites and brachiopods. During his tenure at Yale, he published a number of short descriptions of the remarkable *Triarthrus* trilobites with appendages (Beecher, 1893a, b, 1895a, 1896), and it was Beecher's reconstruction of *Triarthrus* showing its appendages (**50**) that was reproduced in hundreds of textbooks and other publications and brought fame to the finds from Rome. Beecher prepared the trilobites by rubbing away the soft shale from the pyritized legs and antennae with soft erasers. On his untimely death in 1904, Beecher left many specimens, drawings, and an unfinished monograph on the anatomy and relationships of trilobites. One of his former students, Percy Raymond, became professor at Harvard University and completed the monograph which was published in 1920 (Raymond, 1920).

Beecher believed the trilobite quarry to be worked out, but an attempt was made to find it by John L Cisne in 1969, when he was a student at Yale. He found a badly slumped river bank but did not locate any trilobite specimens. He concluded that this may have been the site but that Beecher might indeed have quarried out the localized trilobite bed. Cisne (1973a) produced a detailed palaeoecological study of the trilobite bed based on museum collections. Later, when at Chicago University, Cisne went on to study Beecher's trilobites using X-rays (Cisne, 1975, 1981).

The bed was eventually rediscovered by the meticulous studying of old photographs and published records and careful searching of the creek by Tom Whiteley and Dan Cooper, two industrious fossil collectors, in 1984. Whiteley contacted the landowner, a most helpful local farmer, who brought in a

51 Beecher's quarry today, looking downstream (west).

bulldozer and back-hoe to re-excavate the quarry. It turned out later that their first excavation was actually in Walcott's quarry but, nevertheless, Beecher's Bed had been rediscovered. Once located, further excavations were made in 1989 by a joint team from the Smithsonian Institution, Washington DC, and the American Museum of Natural History, New York (Briggs and Edgecombe, 1992, 1993) and, more recently, by a team from the Yale Peabody Museum. Both Beecher's and Walcott's quarries are now re-exposed for study (**51**), but the main obstacle to collecting lies in the fact that the near-horizontal bed runs straight into the river bank, so tons of overburden would need to be removed to expose more of it.

52 A pyritized *Triarthrus* for which Beecher's Trilobite Bed is famous. This specimen, or one very similar, was the model for **50** (YPM). Animal is about 3 cm (1.2 in) long (excluding antennae).

Stratigraphic setting and taphonomy of Beecher's Trilobite Bed

Beecher's Trilobite Bed is a 4 cm (1.5 in) bed of hard mudrock almost indistinguishable from the under- and overlying Frankfort Shale beds except for its inclusion of beautiful pyritized trilobites (**52**). The surrounding beds contain thin layers rich in the planktonic graptolite *Climacograptus*, scattered parts of *Triarthrus*, articulated brachiopods, and large, orthocone (straight) nautiloids (**53**). There is some post-compaction pyritization of the nautiloids, but little pyrite elsewhere outside the trilobite bed (Briggs *et al.*, 1991). These graptolitic horizons probably represent long periods of low sedimentation during which time the remains of planktonic (floating, e.g. graptolites) and nektonic (swimming, e.g. nautiloids) animals accumulated on the sea floor. It is thought that the pyrite accumulated in the nautiloids because there was still some residual, undecayed organic matter present in their chambers which triggered the precipitation of pyrite. The graptolites, on the other hand, decayed high in the water, so there was little soft tissue left to decompose by the time they landed on the sea floor (Briggs *et al.*, 1991).

The trilobite body parts which are found in the shales outside the trilobite bed are most likely molted exoskeletons, so also lack soft tissue, but the *Triarthrus* specimens in the trilobite bed are complete, not molts. Near the base of the trilobite bed itself is the so-called trilobite layer (Cisne, 1973a) (**54**) which contains a

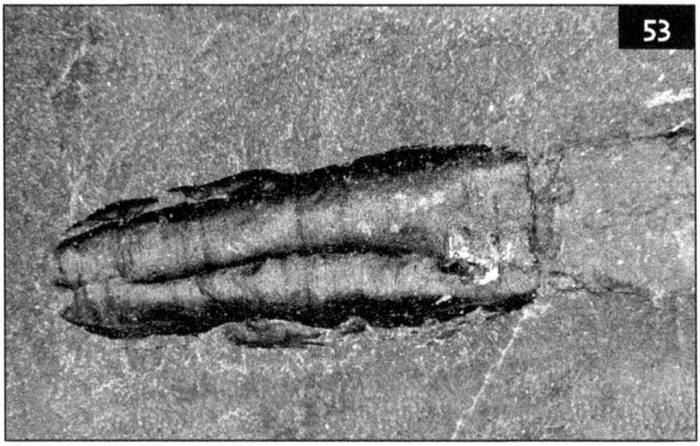

53 An orthocone nautiloid from the Walcott Quarry (YPM). Animal is about 2 cm (0.8 in) long.

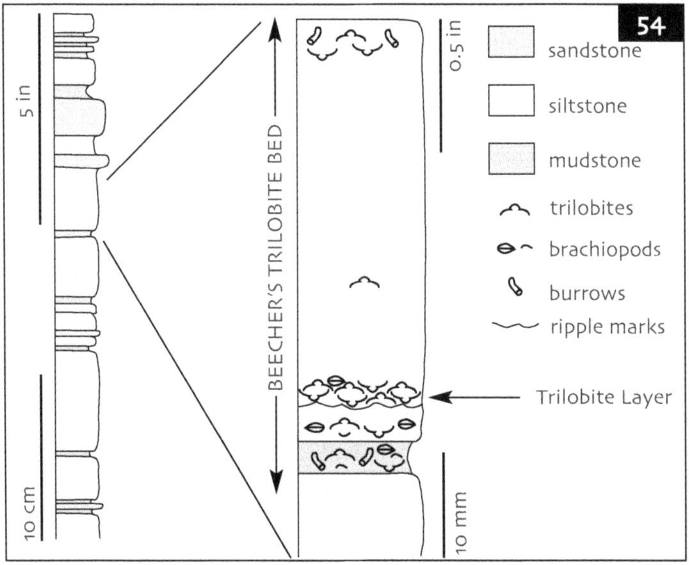

54 Stratigraphic section of Beecher's Quarry (after Bottjer *et al.*, 2001).

dense accumulation of trilobites, brachiopods, and both benthic (bottom-living) dendroid and planktonic graptoloid graptolites, but no orthocone nautiloids. The fauna has been compressed by compaction of the shale, but the the trilobites remained three-dimensional enough while pyritization occurred to preserve them close to their original shape in life. The dorsal exoskeleton and appendages of *Triarthrus* are pyritized, but not the ventral body cuticle (Whittington and Almond, 1987). The compaction has rotated the limbs away from their position in life, and while all authors since Beecher have recognized the importance of compression in the taphonomy of the trilobites, each has imposed a different degree of retrodeformation in their reconstruction: Beecher (1893–1896) unsquashed *Triarthrus* the least, Cisne (1974, 1981) recognized and took it into account somewhat, but Whittington and Almond (1987) used photographs of three-dimensional enrolled specimens of *Triarthrus beckii*, a close relative of *T. eatoni* (see below), published by Ross (1979) to get a truer picture of the convexity of the trilobite in life.

Briggs *et al.* (1991) studied the taphonomy of the trilobite bed because the preservation of soft parts by pyrite is a rare occurrence in the fossil record. They took samples from freshly exposed rock surfaces along a continuous sequence from about 20 cm (approx. 8 in) below to about 15 cm (6 in) above the trilobite bed (**54**) and measured the pyrite sulfur, reactive iron, and organic carbon content. In order to determine the relative timing of formation of the pyrite, the ratio of sulfur isotopes was measured. During the reduction of sulfate by microbes, ^{32}S is reduced more rapidly than ^{34}S. Therefore, sulfate dissolved in water is richer in the lighter isotope, ^{32}S, than the heavier ^{34}S, and is preferentially incorporated into pyrite (FeS_2) until the pore waters in the sediment become distant from the open water, e.g. by further accumulation of sediment. Once the lighter isotope is unavailable, the sulfate-reducing bacteria use the heavier isotope. So, the ratio of the two isotopes (expressed as $\delta^{34}S$) indicates whether the system was open to diffusion or closed, and therefore the relative timing of pyrite formation. The trilobite bed showed the highest level of the heavier isotope (+30.7%) in the sequence studied, indicating that pyrite precipitation continued there longer than elsewhere.

In order for soft parts to be replaced by iron pyrite there needs to be a strong contrast between the chemistry in the region of the carcass and that of the surrounding sediment. Briggs *et al.* (1991) measured low organic carbon concentrations throughout the sequence while the amount of reactive iron was unusually high. Thus, there was insufficient organic matter to feed sulfate-reducing microbes and, indeed, the presence of burrows in the sediments above and below the trilobite bed and the high reactive iron content shows that the sea floor was well oxygenated. So, because of the low organic matter content, little pyrite was formed until the influx of the trilobite bed. This brought in loads of dead trilobites which provided plenty of organic matter for sulfate-reducing bacteria to cause the plentiful iron to diffuse towards the carcasses, react with sulfur, and produce iron pyrite. $\delta^{34}S$ values for the dorsal exoskeleton and the limbs of the trilobites differed in that those of the exoskeleton were always lower in all specimens of *Triarthrus* measured (Briggs *et al.*, 1991). This was because dissolution of the calcium carbonate in the dorsal exoskeleton enhanced the production of pyrite. The pyritization process continued while the sediment was being compacted but eventually stopped and the sediment compacted further. It is likely to have been a matter of some months before pyritization ceased.

Description of Beecher's Trilobite Bed biota

Triarthrus

By far the commonest trilobite in Beecher's Bed is *Triarthrus eatoni* (Hall, 1838) (**55, 56**). The Beecher's Bed specimens were originally identified as *Triarthrus beckii* Green, 1832 by Matthew, Walcott, Beecher, and Raymond, but New York State Palaeontologist Rudolf Ruedemann (1926) and later authors agreed it is really *T. eatoni. T. eatoni* and *T. beckii* can be differentiated by the shape of the palpebral lobe and possibly the glabella (**50**). Also, *T. beckii* occurs mainly in the lowest part of the Upper Ordovician, but is replaced by *T. eatoni* in higher strata of the Upper Ordovician. Indeed, they have been used in biostratigraphy: geologists who encounter Ordovician shales in New York can determine their stratigraphic position by identifying the species of *Triarthrus* which

55 Pyritized *Triarthrus* seen from the ventral side, showing the hypostome between the bases of the antennae, biramous appendages on cephalic, thoracic, and pygidial segments, and note the little, stout spines on some podomeres (YPM). Animal is about 3 cm (1.2 in) long (excluding antennae).

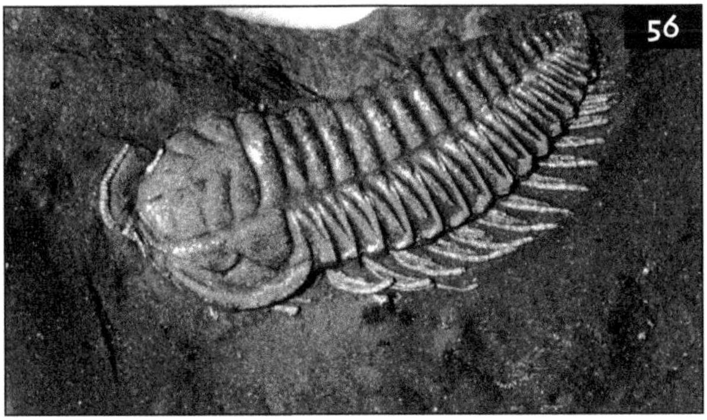

56 *Triarthrus* in an apparent side view, showing the biramous appendages emerging from beneath the dorsal exoskeleton (YPM). Animal is about 3 cm (1.2 in) long (excluding antennae).

they contain. However, Cisne *et al.* (1980) showed, in a detailed morphometric study of 639 specimens of *Triarthrus* through the Trenton Group, that the two species actually grade into one another. The fossils show evidence of an evolutionary cline: the two end-members of the cline are demonstrably different species, but any attempt to identify an intermediate is difficult.

For this reason, Cisne *et al.* (1980) considered the two morphs really to be one species, called *Triarthrus beckii* because this was the earlier of the two names to be given to the fossil, by Green (1832). Ludvigsen and Tuffnell (1983) re-examined the *Triarthrus* problem and came to the conclusion that the two really are distinct species, distinguishable on the basis of the shape of the palpebral lobe. There is some evidence that *T. eatoni* preferred deep-water muddy environments while *T. beckii* occurs more commonly in rocks which represent shallower water environments (Cisne *et al.*, 1980; Whittington and Almond, 1987).

Trilobites have three tagmata (main body parts) called, from front to back: cephalon, thorax, and pygidium (**50**), but note that the three lobes which give these animals their name – trilobite – refer to the median axial and lateral, pleural lobes (**50**). The single pair of antennae are long and consist of many segments which taper gradually to the distal end. The early descriptions by Beecher (1895a, 1896) showed four pairs of biramous (two-branched) limbs beneath the cephalon (**57**). This leg count was followed by Raymond (1920) and Størmer (1951), but Cisne (1974) questioned this number, and his X-ray investigations (Cisne, 1981; **57**) showed only three. The study by Whittington and Almond (1987) confirmed that three was the correct number. Each thoracic segment bears a pair of appendages underneath. Beecher's studies showed the thoracic appendages to consist of two branches, like those of the cephalon. The basal segment (podomere) is the coxa; this segment is attached to the body, though its connection is not clear in any specimen studied, and from the coxa arise the two limb branches. Raymond (1920, Figure 10) suggested that the coxae were shown in this diagram to be too far apart, and this has been confirmed by the studies of Whittington and Almond (1987). The coxae bear spines on their inner side, which, by acting one against the opposite coxa as a pair, are able to push food forwards towards the mouth. In front of the mouth is a plate called the hypostome, which serves to prevent food passing beyond the mouth, though this structure is poorly known in *Triarthrus*. The inner branch (leg branch or endopodite) consists of six podomeres following the coxa; the proximal ones bear downward-pointing spines, and there are

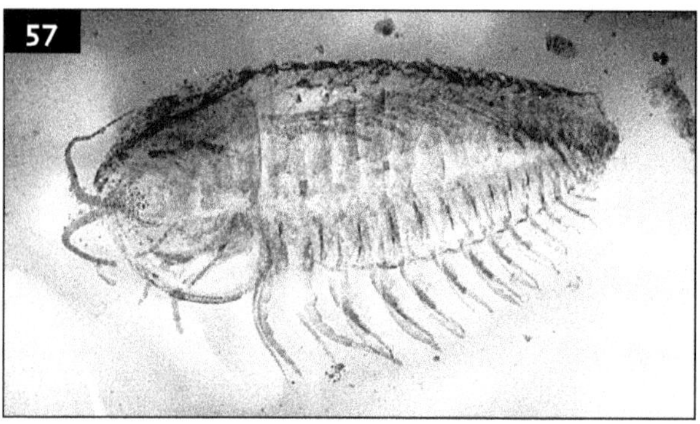

57 X-ray photograph of a *Triarthrus* taken by John L Cisne, preserved in a similar position to the one in **56**. Note the biramous appendages; the appendages on the far side can be seen also (YPM). Animal is about 3 cm (1.2 in) long (excluding antennae).

claws on the end of the most distal podomere. The outer, gill, branch (exopodite) consists of many short podomeres, each supporting a filament, so that the whole structure resembles a gill raker. Almost certainly the inner branch was used for walking, the outer for respiration, but there is some discussion concerning whether the gill branch could also have been used for filter feeding, at least in some trilobites. The appendages get smaller and smaller down the thorax until, beneath the pygidium, they are tiny and must have been useless in locomotion.

Other trilobites

Other trilobites are rare in the trilobite bed, but specimens of *Cryptolithus bellulus, Stenoblepharum beecheri* and a few odontopleurids are known (Whiteley, 1998). *Cryptolithus* (**58**) is a trinucleid; unlike *Triarthrus* these trilobites have no eyes. They have long genal spines, few thoracic segments, and a triangular pygidium. The most extraordinary feature of trinucleids is the fringe around the front of the relatively large cephalon which bears grooves and pits. It is generally considered that trinucleids were bottom-dwelling (benthic) animals which used their long spines as an aid to supporting themselves on the sediment surface. The pits in the fringe are likely to have been sensory: useful in a world where light was probably very dim.

Like all arthropods, trilobites need to molt their cuticles when they grow from juvenile to adult. So, for any adult trilobite, there might be up to a dozen molts left behind as potential fossils. Beecher (1895b) studied the juvenile stages of trilobites, known as protaspids, and found many of these in Beecher's Bed (**59**). He concluded that they must belong to the commonest adult, *Triarthrus.* Whittington (1957) questioned this assumption but Cisne (1973b) agreed with Beecher that the protaspids belonged to *Triarthrus* and produced an interesting analysis of the life stages of the trilobite. Illustrating his study

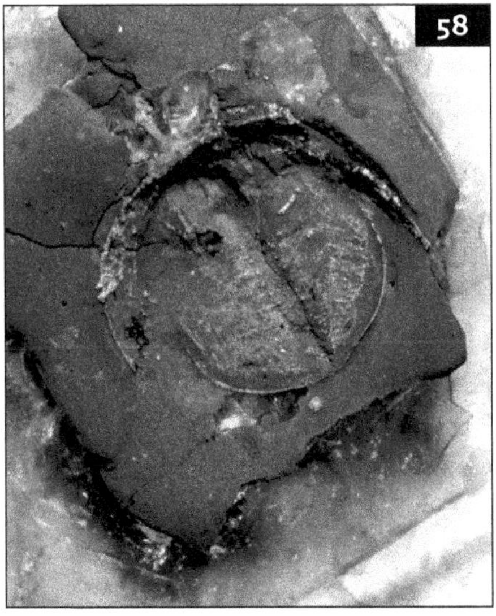

58 A pyritized *Cryptolithus*, ventral view showing thoracic appendages (YPM). Animal is about 2 cm (0.8 in) long.

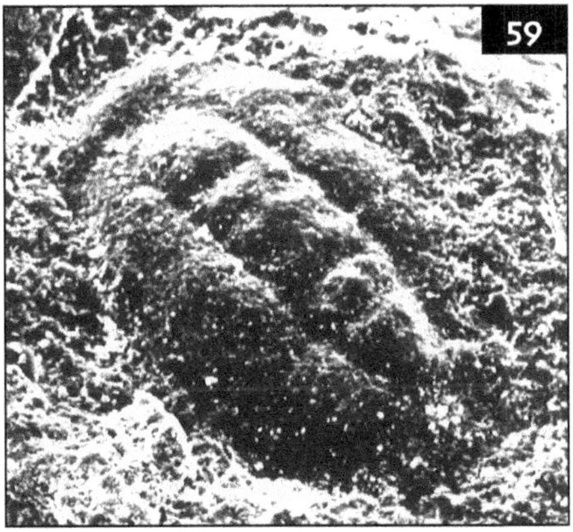

59 Scanning electron microscope photograph of a protaspid (early stage trilobite) of *Stenoblepharum*, showing cephalon, pygidium, and about three thoracic segments (YPM). Animal is about 1.5 mm (0.06 in) long.

with size-frequency distribution graphs, Cisne concluded that the whole population was preserved but the juveniles were more strongly represented because of higher juvenile mortality than deaths in adulthood. However, more recent studies made by Briggs and Edgecombe (1992, 1993) have shown that the protaspids do not belong to *Triarthrus* but to *Stenoblepharum*, adult remains of which are extremely rare in Beecher's Trilobite Bed. So, it appears that *Triarthrus* protaspids lived elsewhere, perhaps in the plankton, but the protaspids of *Stenoblepharum* were benthic while the adults of this trilobite were not.

Other fossils

The trilobites in Beecher's Bed come mostly from the lower few millimetres of the bed, called the mud assemblage; the upper part of the bed contains the silt assemblage which has much rarer trilobites but a few echinoderms. Graptolites (**60**) and brachiopods (**61**) are present in both assemblages.

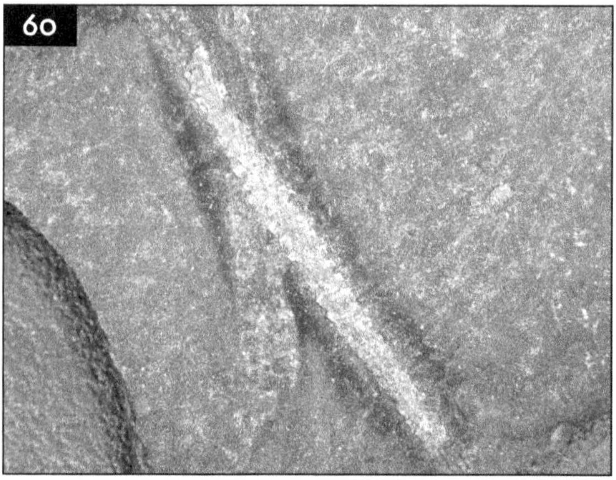

60 The graptolite *Climacograptus*. Note the double-row of thecae, one on each side of the stipe, each of which would have borne an individual polyp-like animal in life (YPM). Animal is about 2.5 cm (1 in) long.

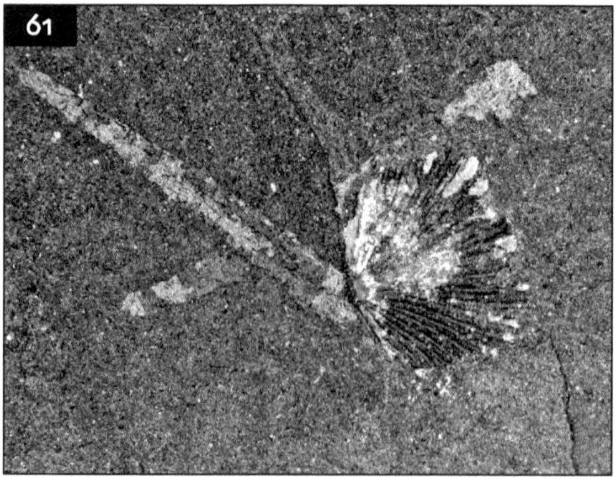

61 Graptolite and orthid brachiopod (YPM). Brachiopod is about 6 mm (0.2 in) wide.

Paleoecology of Beecher's Trilobite Bed

Cisne (1973) made a detailed paleoecological study of Beecher's Trilobite Bed. He made thin sections of the bed itself and the sediments immediately below, and studied the sedimentary structures and fauna (**54**). He showed that immediately below the 40 mm (1.5 in) trilobite bed was a 30 mm (1.25 in) layer of mudstone with burrows, brachiopods, trilobites, and abundant graptolites, which itself overlays a siltstone. The trilobite bed consists of a thin basal layer of coarse silt which shows a sharp lower contact, which then grades gradually into the finer mudrock of the rest of the trilobite bed. The basal layer is followed by a 40 mm (1.5 in) laminated layer with some brachiopods and trilobites, and which has a rippled surface. The main trilobite-rich layer within the trilobite bed is about 2 mm (0.08 in) thick and has no sedimentary or biogenic structures but is packed with brachiopod, trilobite, and graptolite fossils. Most of the rest of the trilobite bed has very rare trilobites, except near the top layer, which has more fossils as well as burrows (**62**).

Cisne (1973) concluded that the trilobite bed represented a micro-turbidite. Turbidites are packets of sediment which flow down an underwater slope, e.g. the front of a delta, into deeper water by gravity, and may be triggered by an earthquake or similar event. They are similar to avalanches on land. The coarse detritus in the turbidite flow settles out first, and nearest (proximal) to the origin of the flow, whereas finer particles can spread for many miles as a wide apron of thin flow consisting of silt and mud-sized particles. As they flow, turbidites erode and scour the sea floor, leaving behind tell-tale sole marks such as flute casts on the base of their bed. When they come to a stop, coarser particles settle out first and then finer and finer muds. For this reason they show grading from a sharp, erosional base up into finer muds at the top. Beecher's Trilobite Bed

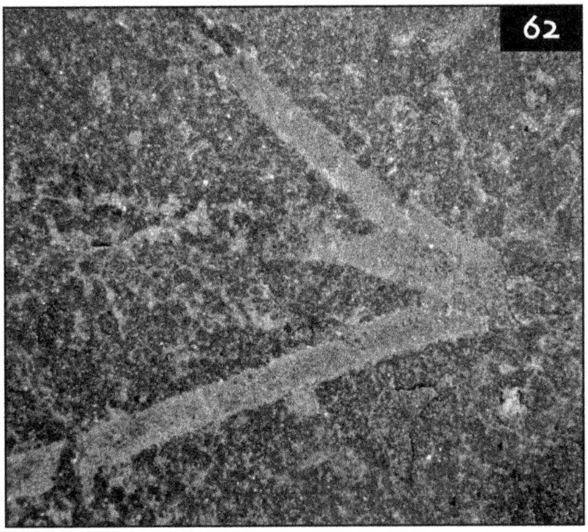

62 A three-branched burrow (YPM). Length of longest branch about 3 cm (1.2 in).

shows these features, as well as internal ripples and laminae which are known in microturbidites (**63**).

Cisne (1973) studied the orientation of the fossils in the trilobite bed (**64**) and showed that the carcasses were equally divided between those which lay dorsal-side up and those ventral-up. The carcasses were strongly aligned in a single direction: the long axes of the trilobites were mostly parallel to each other, but the head could be oriented either way (**64**). This means that the trilobites were laid down in a current which oriented them parallel to its direction of flow. Being dead animals, there was no difference in stability whether they were dorsal- or ventral-up; had they been only dorsal exoskeletons, they would have been more stable dorsal-side up (i.e. more easily flipped over if ventral-up) and more easily disarticulated.

So we can envisage the mud assemblage fauna consisting of benthic trilobites, with *Triarthrus* as the commonest, some *Cryptolithus*, and protaspids of *Stenoblepharum*. *Triarthrus* most likely fed on organic matter on the sea floor: a scavenger and predator, digging into the mud and passing food forward with its coxae to the mouth at the front. *Cryptolithus*, too, was a bottom-feeder. In the water column there were planktonic graptolites and nektonic orthocone nautiloids, as well as possibly some swim-

63 Cross-section through the trilobite layer showing cross-bedding and a burrow (bottom left) (YPM). Thickness 3 cm (1.2 in).

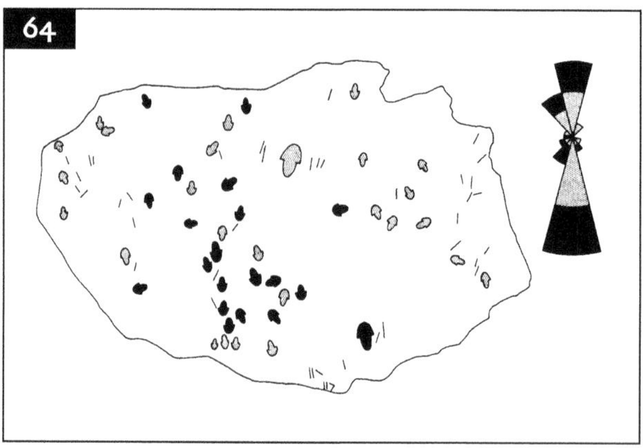

64 Plan view of slab of trilobite layer showing orientation of *Triarthrus* trilobites and *Climacograptus* graptolites. Stippled trilobites indicate dorsal side uppermost, black indicates ventral-up. Rose diagram to right shows the orientation of fossils. Note that there is a preferred orientation (aligned with the top–bottom of the page), and that there is no preferred way-up (about half of the trilobites are ventral-up, half dorsal-up). Average trilobite length about 3 cm (1.2 in). After Cisne (1973a).

ming trilobites such as odontopleurids. When bottom conditions were more oxygenated, soft-bodied worms burrowed into the sediment and left their traces behind. So, it seems that the microturbidite disturbed this happy scene and transported the live animals in a single event from their original feeding site just a short distance to where they were deposited entombed in mud. What killed the animals could have been smothering with sediment (i.e. lack of oxygen), or temperature shock: while the bottom waters where the animals lived was likely to have been cold, it might have been colder still where they landed (cf. the Burgess Shale, Chapter 3).

Triarthrus belongs to the trilobite family Olenidae, characterized by thin cuticles, wide pleural regions, and many thoracic segments (hence many appendages). Their abundance in dark mudstones, rich in sulfides, suggests they were tolerant of low-oxygen conditions. This led Fortey (2000) to suggest that they could utilize sulfur bacteria as chemo-autotrophic symbionts. These bacteria live in conditions of low oxygen, often just at the interface between oxic and anoxic environments, and use sulfur (commonly as hydrogen sulfide) as a source of energy. Fortey suggested that the peculiar morphology of the later olenids (*Triarthrus* is among the last of the family) is consistent with them utilizing chemo-autotrophic bacteria. *Triarthrus* has many appendages with gill filaments which could act as substrates for bacterial cultivation; the lack of a hypostome in some olenids (it is poorly known in *Triarthrus*) is consistent with it being unnecessary if food was absorbed directly through the gill filaments. This method of feeding is known from a wide variety of organisms which inhabit hypoxic (low oxygen) environments today; examples are known from crustaceans, worms. and bivalve mollusks, which show varying degrees of adaptation to bacterial cultivation. If Fortey's idea is correct, then olenids are the oldest animals known to have used this feeding method.

Comparison of Beecher's Trilobite Bed with other Lower Paleozoic biotas

Beecher's Trilobite Bed is just the thin (40 mm, 1.5 in) microturbidite bed; the other Frankfort Shale layers above and below, which also include microturbidites but lack the extraordinary pyrite preservation, contain similar faunas, and are typical of Ordovician deep-water faunas. There are few other Ordovician Lagerstätten known. The Soom Shale in South Africa (Selden and Nudds, 2004, Chapter 3) preserves a unique fauna from a high-latitude (60°S), cool, glacially influenced marine habitat. The Soom Shale contains trilobites, including a trinucleid, but also has a strange naraoiid similar to *Naraoia* of the Burgess Shale (Chapter 3). Both the Soom Shale and Beecher's bed yield have orthocone nautiloids, and these large animals seem to have been common members of the nekton throughout the Ordovician Period. Inarticulate brachiopods are also found in both Lagerstätten. The Soom Shale has the eurypterid *Onychopterella*, which is not known from the Frankfort Shale but occurs in Silurian beds such as the Kokomo Limestone of Indiana. Both the Soom Shale and Beecher's Bed yield belong to the late Ordovician, Ashgill Epoch. At this time, Earth experienced global cooling, with a glaciation centred on the part of Gondwana now occupied by north and west Africa. Its effects were wide-ranging, and in New York there was a noticeable severe decline in the diversity of graptolites, plankton which are normally associated with low, subtropical latitudes. This end-Ordovician mass extinction event wiped out nearly a quarter of animal families living at the time. It was the greatest extinction to affect trilobites during their time on Earth; more than half of all trilobite families, including the olenids and trinucleids, did not survive to the Silurian Period.

Further Reading

Bartels, C., Briggs, D. E. G. and Brassel, G. 1998. *Fossils of the Hunsrück Slate – marine life in the Devonian.* Cambridge University Press, Cambridge.

Beecher, C. E. 1893. On the thoracic legs of *Triarthrus. American Journal of Science,* 3rd series **46**, 467–470.

Beecher, C. E. 1894a. On the mode of occurrence, and the structure and development of *Triarthrus beckii. American Geologist* **13**, 38–43.

Beecher, C. E. 1894b. The appendages and pygidium of *Triarthrus. American Journal of Science,* 3rd series **47**, 298–300.

Beecher, C. E. 1895a. Further observations on the ventral structure of *Triarthrus. American Geologist* **15**, 91–100.

Beecher, C. E. 1895b. The larval stages of trilobites. *American Geologist* **16**, 166–197.

Beecher, C. E. 1896. The morphology of *Triarthrus. American Journal of Science,* 4th series **1**, 251–256.

Bottjer, D. J., Etter. W., Hagadorn, J. W. and Tang, C. M. (eds.) 2002. *Exceptional Fossil Preservation.* Columbia University Press, New York.

Briggs, D. E. G., Bottrell, S. H. and Raiswell, R. 1991. Pyritization of soft-bodied fossils: Beecher's Trilobite Bed, Upper Ordovician, New York State. *Geology* **19**, 1221–1224.

Briggs, D. E. G. and Edgecombe, G. D. 1992. The gold bugs. *Natural History* **101**, 36–43.

Briggs, D. E. G. and Edgecombe, G. D. 1993. Beecher's Trilobite Bed. *Geology Today* **1993**, 97–102.

Cisne, J. L. 1973a. Beecher's trilobite bed revisited: ecology of an Ordovician deep water fauna. *Postilla* **160**, 1–25.

Cisne, J. L. 1973b. Life history of an Ordovician trilobite *Triarthrus eatoni. Ecology* **54**, 135–142.

Cisne, J. L. 1974. Trilobites and the origin of arthropods. *Science* **186**, 13–18.

Cisne, J. L. 1975. Anatomy of *Triarthrus* and the relationships of the Trilobita. *Fossils and Strata* **4**, 45–63.

Cisne, J. L. 1981. *Triarthrus eatoni* (Trilobita): anatomy of its exoskeletal, skeletomuscular, and digestive systems. *Palaeontographica Americana* **9**, 99–142.

Cisne, J. L., Molenock, J. and Rabe, B. D. 1980. Evolution in a cline: the trilobite *Triarthrus* along an Ordovician depth gradient. *Lethaia* **13**, 47–59.

Fortey, R. A. 2000. Olenid trilobites: the oldest known chemoautotrophic symbionts? *Proceedings of the National Academy of Sciences of the USA* **97**, 6574–6578.

Green, J. 1832. *A Monograph of the Trilobites of North America.* Joseph Brano, Philadelphia.

Hall, J. 1838. Descriptions of two species of trilobites belonging to the genus *Paradoxides. American Journal of Science* **33**, 137–142.

Ludvigsen, R. and Tuffnell, P. A. 1983. A revision of the Ordovician olenid trilobite *Triarthrus* Green. *Geological Magazine* **120**, 567–577.

Matthew, W. D. 1893. On antennae and other appendages of *Triarthrus beckii. American Journal of Science,* 3rd series **46**, 21–125.

Raymond, P. E. 1920. The appendages, anatomy and relationships of trilobites. *Memoirs of the Connecticut Academy of Arts and Sciences* **7**, 1–169.

Ruedemann, R. 1926. The Utica and Lorraine formations of New York. Part 2. Systematic paleontology. No. 2, mollusks, crustaceans, and eurypterids. *New York State Museum Bulletin* **272**, 1–227.

Selden, P. and Nudds, J. 2004. *Evolution of Fossil Ecosystems.* Manson, London.

Sepkoski, J. J. 1979. A kinetic model of Phanerozoic taxonomic diversity. II. Early Phanerozoic families and multiple equilibria. *Paleobiology* **5**, 222–251.

Størmer, L. 1951. Studies on trilobite morphology. Part III. The ventral cephalic structures with remarks on the zoological position of the trilobites. *Norsk Geologisk Tidsskrift* **29**,108–158.

Walcott, C. D. 1876. Preliminary notice of the discovery of the remains of the natatory and branchial appendages of trilobites. *28th Report of the New York State Museum of Natural History*, advance sheets, 89–91.

Walcott, C. D. 1879. Notes on some sections of trilobites from the Trenton Limestone. *31st Report of the New York State Museum of Natural History*, 2–10.

Walcott, C. D. 1894. Notes on some appendages of the trilobites. *Proceedings of the Biological Society of Washington* **9**, 89–97.

Whiteley, T. E. 1998. Fossil Lagerstätten of New York, Part 1: Beecher's Trilobite Bed. *American Paleontologist* **6**, 2–4.

Whittington, H. B. 1957. Ontogeny of *Elliptocephala, Paradoxides, Sao, Blainia*, and *Triarthrus* (Trilobita). *Journal of Paleontology* **31**, 934–946.

Whittington, H. B. and Almond, J. E. 1986. Appendages and habits of the Upper Ordovician trilobite *Triarthrus eatoni. Philosophical Transactions of the Royal Society of London*, series B **317**, 1–46.

The Bertie Waterlime

Chapter 5

Background

At the end of the Ordovician Period there was a major extinction event which coincided with a worldwide glaciation and, in New York State, with a major erosional episode that produced the Cherokee Unconformity at the Ordovician–Silurian boundary. The Silurian Period in New York begins, therefore, with a marine transgression, as the sea invaded the land following the melting of the ice sheets and the wearing down of the exposed land surface. The Silurian is a relatively shorter period than the Ordovician and, although most of the time sea levels were fairly high, there was a considerable variety of environments which are represented today by different rock types. The sea spread across New York from the west, laying down sandstones and shales derived from erosion of the land to the east. By mid-Silurian times, open sea covered most of the state and shales and sandstones gave way to increasing amounts of limestone. Increased tectonic activity rejuvenated mountains to the east which resulted in the sea floor becoming smothered with sand (in the east), shales (mid-state) and deep-water muds in the west. As the mountains wore down, so quiet-water conditions again prevailed and the later half of the Silurian Period in New York is represented with rock sequences predominantly carbonate in nature. These include limestones, dolomitic limestones (dolostones), and also beds of evaporite (salt, gypsum, and anhydrite). This last sequence of rocks is known as the Salina Group.

Dolostones and evaporites are characteristic of shallow-water, restricted circulation basins: a landlocked sea with an arid climate (New York was about 30° south of the equator at this time). The widespread evaporation caused the formation of primary dolostone (magnesium carbonate-rich limestone) and the growth of crystals of halite (rock salt, NaCl), anhydrite ($CaSO_4$) and its hydrated form, gypsum ($CaSO_4.2[H_2O]$).

The period of sedimentation of the Salina Group of dolostones and evaporites was brought to a close by the deposition of the Camillus Shale: a thick accumulation of mottled shales with halite and gypsum crystals, an important source of these minerals in western New York. But the rock sequence we are most interested in overlies the Camillus Shale. The highest beds in the Silurian of New York are the Bertie and Rondout groups. The Bertie consists of a muddy dolostone having the peculiar and useful characteristic of being able to be turned into cement which will harden under water; hence it was given the name Waterlime. Arid, evaporating conditions continued through the Bertie Waterlime times, and some horizons in the rocks of this group contain the abundant remains of unusual animals and land plants, indicating an input from nearby rivers. At the very end of the Silurian Period in New York, the Rondout Group marks the return to more normal marine salinities and fossils similar to those of the mid-Silurian.

The paleobiologic importance of the Bertie Waterlime lies in its abundance of well-preserved remains of eurypterids and other unusual arthropods. Eurypterids belong to the arthropod subphylum Chelicerata, which also contains horseshoe crabs, spiders, scorpions, mites, ticks, harvest-spiders, and other arachnids. They are united in their possession of a characteristic pair of pincers (chelicerae) as the first pair of appendages. Eurypterids arose in the Ordovician and died out in the Permian; some reached some 2 m (6.6 ft) or more in length, and they were predatory. Thus, they were the largest arthropods that ever lived, and were the top predators on Earth for some 100 million years!

History of discovery of the Bertie Waterlime

The first eurypterid ever to be mentioned in the literature was found near Westmoreland, Oneida County, New York, and was described as a catfish of the genus *Silurus* (Mitchill, 1818). Mitchill thought that the prominent swimming paddles were barbels arising from near the mouth. James DeKay (1825) recognized the arthropod nature of the fossil and named it *Eurypterus remipes.* He thought it belonged to the Crustacea. *Eurypterus lacustris,* the common eurypterid in the western part of New York, was discovered soon after, by Richard Harlan (1835). It was more than 20 years later that the first complete eurypterid was discovered outside New York, in Estonia by Nieszkowski (1858, 1859). He considered this species, now known as *Eurypterus tetragonophthalmus,* to be so like the New York *E. remipes* that he placed it in the same species.

Through the nineteenth century, masses of *Eurypterus* were collected from the extensive waterlime quarries and deposited in the museums of New York State. Indeed, apart from a similar fauna of the same age from Kokomo, Indiana, no eurypterids were found in the US outside the state of New York until the twentieth century. Because of the abundance of these beautiful fossils in the Bertie Waterlime of New York, the species *Eurypterus remipes* was designated as the state fossil in June 1984.

Stratigraphic setting and taphonomy of the Bertie Waterlime

The Bertie Group was formerly included within the Salina Group (e.g. Rickard, 1975), but is now generally separated. The Salina Group (**65**) consists primarily of thick red-beds (Vernon and Bloomsburg formations), evaporites, limestones, and dolostones (Syracuse Formation), with an upper, thick sequence of shales with dolomitic mudstones and evaporites (Camillus Formation). In contrast, the Bertie Group is a thinner sequence, consisting of massive dolostones with intercalated waterlimes, minor shale and mudstone units, and some evaporites (Ciurca and Hamell, 1994). Eurypterids occur in specific horizons within the Salina Group, such as the Vernon Shales, but are unknown in the Camillus Formation.

The Bertie Group overlying the Salina has some massive dolostone horizons which are good waterfall-formers, e.g. Indian Falls, just west of NY77 (**66**) and the falls at Williamsville, near Buffalo. The name Bertie was coined by Chapman (1864) for rocks in the neighborhood of the town of Bertie in the Niagara Peninsula, Ontario. Judging from the thickness given by Chapman, he must have included the strata all the way up to the basal Devonian unconformity in the Bertie (Ciurca and Hamell, 1994), whereas more recently, the Akron Formation and its equivalent Cobleskill Formation to the east have been included

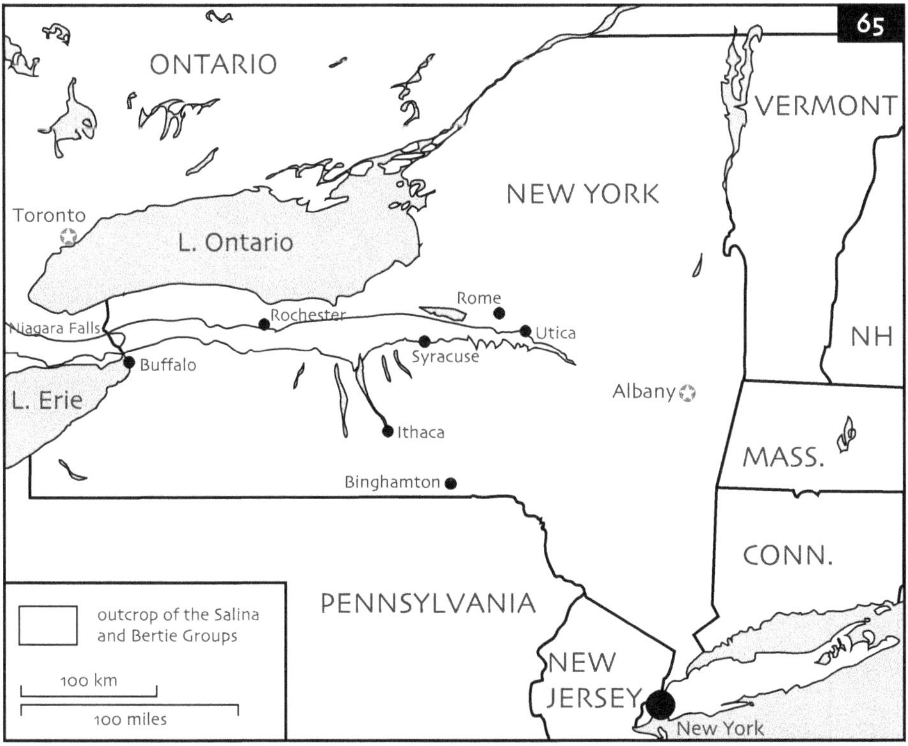

65 Map of New York State and southern Ontario showing the extent of the Salina and Bertie Groups of the Silurian sequence.

66 Indian Falls near Pembroke, Genesee County, New York, is formed where Tonawanda Creek drops over massive beds of the Victor Dolostone overlying the Morganville Member (Fiddlers Green Formation).

in the Rondout Group (**67**). However, discovery of a eurypterid-bearing waterlime above the Akron Formation: the Moran Corner Waterlime (Ciurca, 1990), prompted Ciurca and Hamell (1994) to follow Chapman and include the Akron–Cobleskill and the Moran Corner in the Bertie Group.

The Bertie Group sediments record a cyclic sequence of transgressions

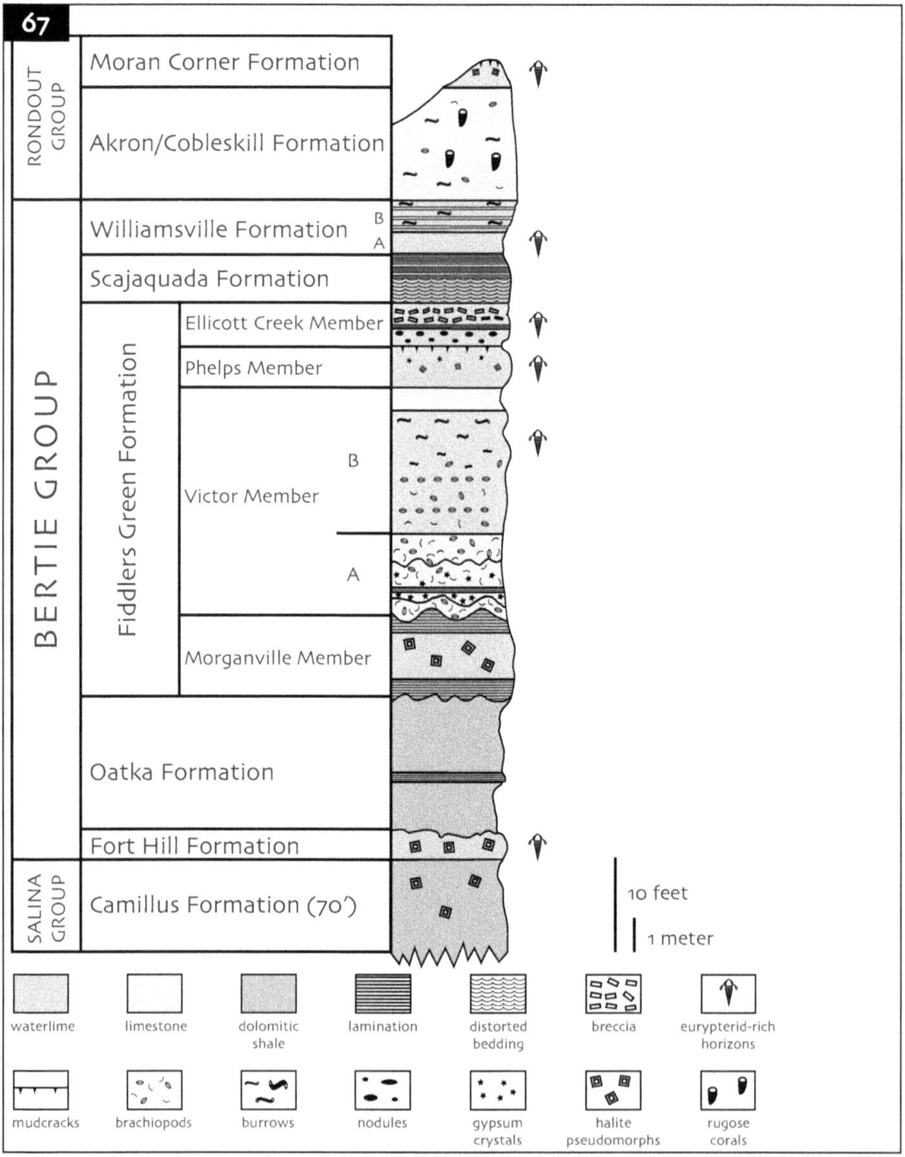

67 Stratigraphic section of the Bertie and Rondout Groups in western New York State. Based on Ciurca and Hammell (1994).

and regressions of the sea across an intertidal-subtidal platform. The sequence represents a range of environments from brackish estuarine, through shallow evaporitic and sabkha settings, to near-normal marine lagoonal and subtidal conditions. Two formations are of especial importance for fossils: the lower Fiddlers Green Formation and the upper Williamsville Formation. (**67**) shows subdivisions of the Fiddlers Green into a number of members, which indicates that there is considerable variation of environments, even within the waterlime. Laminations suggest cyclic sedimentation on a seasonal scale, and there is strong evidence of algal (stromatolite) structures. In some places, algal mounds can be seen, and eurypterids are common at these locations. Also of great interest are the giant salt pseudomorphs (**68**). Rock salt (halite, NaCl) has a cubic crystal structure and habit. Commonly, the former presence of salt crystals can only be inferred by pseudomorphs, where the crystal has been dissolved by an influx of water followed by deposition of sediment into the mold where the crystal has been. Many of the pseudomorphs show hopper-faces, in which the normally flat faces of the cubic crystal were hollow and bore stairway-like stepped sides.

Environments evidenced by the rock types include the following: (1) Supratidal sabkha: a barren, hypersaline plain above normal high tide mark but subject to periodic flooding; examples of this environment are the dolomitic shales of the Camillus, Oatka and Scajaquada formations. (2) Supratidal storm-generated breccias occur in the Ellicott Creek Member of the Fiddlers Green Formation. (3) Hypersaline lakes and estuaries are represented by massive gypsum beds. (4) The intertidal environment is the one in which the massive waterlimes of the Fort Hill Formation and the Morganville and Phelps members of the Fiddlers Green Formation were deposited. Regular exposure and evaporation led to the development of large hopper salt crystals. Algal mats and eurypterids are common, but occasional ghosts of nautiloids and gastropods suggest a much greater diversity of organisms with calcareous shells which are leached and so poorly represented in the fossil biota. (5) The subtidal environment preserved a greater proportion of shelly faunas, such as the

68 Giant salt hopper pseudomorph, Ellicott Creek Member (Fiddlers Green Formation), floor of Neid Road Quarry, Le Roy, New York.

brachiopod *Whitfieldella* and the ostracode *Leperditia*. Trace fossils are also more abundant in this environment, which is represented by the Victor Member of the Fiddlers Green Formation.

The superb preservation of the non-biomineralized biota in this Konservat-Lagerstätte has been attributed to rapid burial of their remains in very fine carbonate muds with a lack of benthic scavengers and bioturbators due to a combination of periodically elevated salinities and bottom anoxia (Kluessendorf and Mikulic, 1991). These authors suggested that the eurypterid remains were introduced into the lagoon as molts (exuviae), rather than live or dead animals, and therefore the anoxia was responsible for the preservation of the remains rather than death of the animals.

Description of the Bertie Waterlime biota

Eurypterids

Commonly known as sea scorpions, the body of eurypterids (**69**, **70**) is divided into

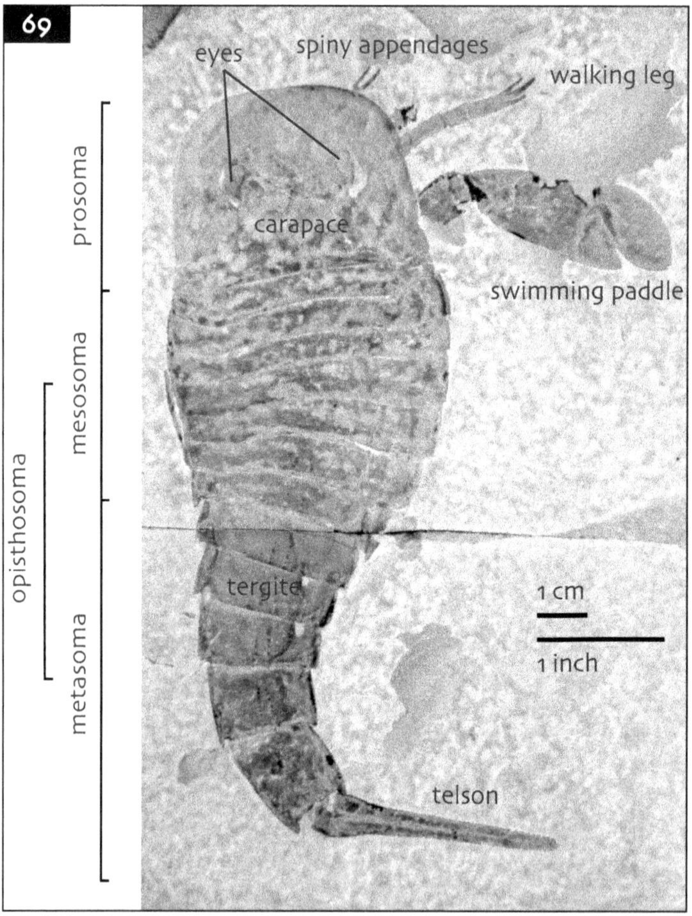

69 *Eurypterus lacustris*, Williamsville Formation, Ridgemount Quarries, Ontario (YPM). Annotated to show the dorsal morphology of eurypterids.

two parts: the anterior prosoma and the posterior opisthosoma. The prosoma bears a carapace with a pair of compound eyes and six appendages. The first appendage is the chelicera, which is usually small, but can be greatly enlarged in the pterygotids. The following three pairs of appendages are usually spiny and used for food capture, e.g. in *Eurypterus*, and increase in size backwards but, in pterygotids, which use their enlarged chelicerae for prey capture, these appendages bear no spines. The fifth appendage is usually leg-like in all eurypterids, and the sixth is normally adapted as a swimming paddle, although it is leg-like in the stylonurids which presumably walked rather than swam. The posterior part of the body, the opisthosoma is itself divided into an anterior mesosoma of seven segments, which bears gills beneath flaps on the underneath, and a posterior metasoma of five segments plus a telson which can be a spine (e.g. in *Eurypterus*), a possible stinger (e.g. in carcinosomatids), or a flipper (in pterygotids).

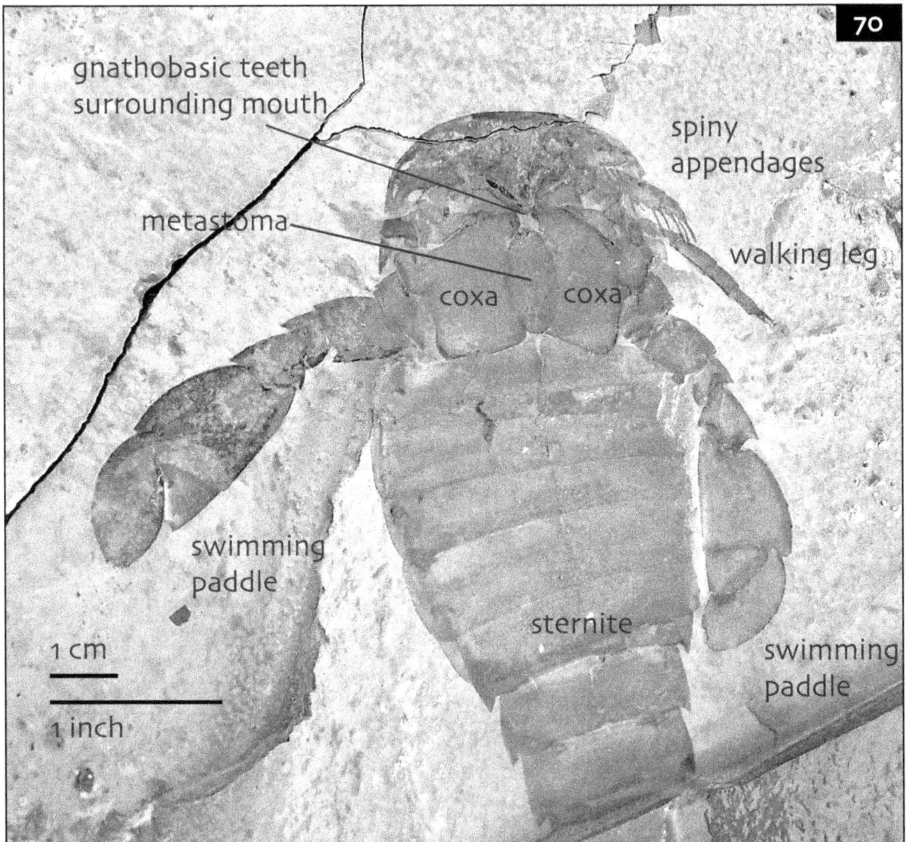

70 *Eurypterus lacustris*, Williamsville Formation, Ridgemount Quarries, Ontario (YPM). Annotated to show the ventral morphology of eurypterids.

There are essentially two, somewhat different, eurypterid faunas in the Bertie Group. The Fiddlers Green Formation is best developed in the eastern part of the outcrop, especially at the well-known Passage Gulf locality. The Williamsville Formation is better developed in the western part of the outcrop, around Williamsville itself (near Buffalo, **71**) and in quarries in Ontario, such as the Ridgemount complex (**72**). However, the horizon and locality from which specimens came from were not always recorded. Early in the history of studies on the eurypterids of New York, it was apparent that there were two forms of *Eurypterus* which formed the bulk of specimens collected: *Eurypterus remipes* (**73**) and *E. lacustris*. They can be distinguished by size and the width–length ratio of the carapace: *E. lacustris* is supposedly larger and slimmer than *E. remipes*.

71 Type locality of the Williamsville Formation adjacent to the old mill at Williamsville, New York, and Sam Ciurca, prolific collector of Bertie fossils.

In their monumental monograph on the eurypterids of New York, Clarke and Ruedemann (1912) distinguished the two species and noted that *E. remipes* occurs in the Herkimer County localities of the Bertie Waterlime whereas *E. lacustris* is found in the Erie district (see also Ruedemann, 1925). Kjellesvig-Waering (1958) considered that the differences between these species and the European *E. tetragonophthalmus* were so slight as to be of mere subspecific status. Thus he recognized a single species *Eurypterus remipes* with subspecies *remipes*, *lacustris*, and *tetragonophthalmus*. (Kjellesvig-Waering, 1958, p. 1137). Andrews *et al.* (1974), in the first statistical morphometric study of eurypterids, agreed with the subspecific status of the European and American *E. remipes*. However, Størmer (1973) considered the paddle shape to be sufficiently different to warrant separation of the European species into a different genus, *Baltoeurypterus*. Kjellesvig-Waering later concurred with Størmer's view and, indeed, described a new species of *Baltoeurypterus* (Kjellesvig-Waering, 1999). It should be pointed out that the males of *E. tetragonophthalmus* have a distinctive scimitar lobe on appendage II, which is unknown in any other species of *Eurypterus* (Størmer and Kjellesvig-Waering, 1969; Selden, 1981; Braddy and Dunlop, 1997).

More recent morphometric work by Tollerton (1992) indicated that while *E. lacustris* and *E. remipes* could be told apart by carapace shape at large sizes, they were not distinguishable as juveniles. He surmised that *E. lacustris* (which occurs in the higher Williamsville strata) could have

72 Quarry of the Ridgemount complex, Ontario, in Williamsville Formation, showing specimens of *Eurypterus lacustris* on waterlime slabs in the foreground.

73 *Eurypterus remipes*, Fiddlers Green Formation, Lang's Quarry, Ilion, Herkimer County, New York. (NHM). Total length 20 cm (<8 in).

developed from *E. remipes* (which is found in the lower Fiddlers Green Formation) by peramorphosis: phylogenetic change in which individuals of a species mature past adulthood and develop new traits, i.e., the opposite of pedomorphosis. However, in the study by Cuggy (1994), *lacustris* and *remipes* plotted on the same ontogenetic trend. He concluded that the two belonged to the same species and synonymized them (as *E. remipes*). A recent cladistic study of the *Eurypterus* genus by Tetlie (2006) placed *E. remipes* and *E. lacustris* as sister groups, and *E. tetragonophthalmus* as sister to *E. hennigsmoeni* from Norway (Tetlie, 2002). The important conclusion from Tetlie's (2006) study was that the genus *Baltoeurypterus* was invalid because its members nested within the *Eurypterus* clade.

The most recent study of the subject by Tetlie *et. al* (2007), showed that the problem with older studies was that the provenance of the specimens could not always be assured. These authors obtained large samples of both forms from known horizons and were able to show consistent morphological differences (primarily carapace shape and telson serrations) between *E. lacustris* and *E. remipes* at all sizes. They concluded that the species are genuine, and that *E. lacustris* evolved from *E. remipes* through heterochrony, as originally suggested by Tollerton (1992).

There are other eurypterids known from the Bertie Waterlime, though not in such abundance as those already mentioned. Another *Eurypterus*, *E. dekayi*, can be distinguished from the other species of this genus by its more numerous (4–6) spines on each segment of its spiny prosomal appendages. This species is very similar to *Eurypterus laculatus*, first recognized by Kjellesvig-Waering (1958) (as *E. remipes laculatus*). Both species have a rather broad carapace but *E. laculatus* has a characteristic depression surrounding the eyes and lacks the ornament of scales which occurs at the rear of the carapace and first segment of the mesosoma on other *Eurypterus* species. *Eurypterus dekayi* occurs in the *E. lacustris* Fauna of the Williamsville Formation at localities from Ontario to Rochester. *E. laculatus* occurs widely in the Fiddlers Green Formation from Ontario to eastern New York. It is possible that *E. dekayi* evolved from *E. laculatus*.

Related to *Eurypterus* is the genus *Erieopterus*, which was erected by Kjellesvig-Waering (1958) to distinguish forms with a smooth ornament, a definite constriction of the body between the mesosoma and metasoma, and lack of spines on the anterior appendages. *Erieopterus* occurs in the Devonian strata immediately overlying the Bertie, but not in the Bertie Waterlime itself. One species from the Williamsville Formation, with a pustulose ornament and a bizarre, globular telson, was originally included in *Erieopterus* but then placed in a new genus *Buffalopterus* by Kjellesvig-Waering and Heubusch (1962). Though the whole animal is unknown, fragments of carapace indicate that this species, *Buffalopterus pustulosus*, reached 1 m (3.3 ft) in length! Another group of eurypterids not too distantly related to the family Eurypteridae is the family Dolichopteridae. *Dolichopterus* has larger eyes and more spiny anterior appendages than *Eurypterus*. The most posterior prosomal appendage is a swimming paddle, as usual, but the penultimate appendage is also somewhat flattened and would also have helped in swimming. *Dolichopterus* occurs through-out the Bertie Waterlime.

Carcinosomatids are unusual eurypterids, with a trapazoidal carapace with eyes situated near the front corners, very spiny anterior appendages, large swimming paddles, and a scorpion-like curved telson. These eurypterids were large and their spiny appendages and probably poisonous stinger indicate that carcinosomatids were fearsome predators. *Paracarcinosoma scorpionis* (**74**) occurs in the Williamsville Waterlime from the Syracuse area westwards to Ontario.

Paracarcinosoma probably reached a length of at least 1 m (3.3 ft).

By far the largest eurypterids from the Berte Waterlime, and the biggest eurypterids known from anywhere, are the pterygotids. Being so large, whole pterygotids are rarely found, so we can only make estimates of their size in real-life from fragments. The specimen of *Acutiramus macrophthalmus* shown in (**75**) was discovered in 1964 in the Phelps Member of the Fiddlers Green Formation in eastern New York and is one of the largest arthropod fossils

74 Spiny, food-capturing appendage of *Paracarcinosoma scorpionis*, Williamsville Formation, Ridgemount Quarry, Ontario (YPM). Length 4.3 cm (1.7 in).

75 Sam Ciurca with a specimen of the giant pterygotid *Acutiramus* in 1965.

known. The Williamsville Waterlime bears a related species: *Acutiramus cummingsi*. It is a wider, more robust form. *Acutiramus* gets its name because its distinctive chelicera has acute tips to each finger (**76**). The large size of the animal and the enormous chelicerae with sharp, serrated teeth suggest that this animal was a fearsome predator. The swimming paddles and rudder-like telson tell us that the animal was a swimmer, though its large size suggests that it might not have swam very fast. It can be imagined cruising the Silurian seas, lashing its long claws out at the small, primitive, jawless fish and other animals it preyed upon.

Xiphosurans

Horseshoe crabs are familiar to those who frequent the eastern seashores of North America, where the remains of *Limulus polyphemus* are commonly washed up on beaches and, at certain times of the year, hordes of these animals come ashore and up estuaries to mate by moonlight (see the delightful book by Milne and Milne, 1965). The epithet 'living fossil', given to *Limulus*, results from the remains of animals strikingly similar in morphology that are preserved in the Mesozoic rocks of Solnhofen, Germany (see Selden and Nudds, 2004, Chapter 10), and quite similar forms going back to the Pennsylvanian Period (see Mazon Creek, Chapter 7). Earlier, more primitive, relatives of horseshoe crabs are the socalled synziphosurines, which include two species known from the Bertie Waterlime: *Pseudoniscus clarkei* and *Bunaia woodwardi*. Synziphosurines differ from true horseshoe crabs in that their opisthosomal segments are free, as in eurypterids, not fused into a hard shield or buckler, as in *Limulus*. *Pseudoniscus clarkei* (**77**) was described by Ruedemann (1916); it is the commoner of the two, and most specimens come from the Williamsville Formation bed A of Ontario. A similar species, *P. roosevelti* (Clarke, 1902), occurs in lower beds of the New York Silurian, and some authors consider that the two species are actually the same (Eldredge, 1974). *Bunaia woodwardi* is a small, enigmatic form described by Clarke (1919). Eldredge (1974) took another look at it – there is only one slab with four specimens on, preserved in the New York State Museum. He wondered whether these little creatures might actually be young specimens of *Pseudoniscus*, but we can only continue to wonder until more specimens of, ideally, a whole range of sizes, show a transition between the two genera.

Scorpions

In the space of about 5 years towards the end of the nineteenth century nearly all of the few known Silurian scorpion fossils were discovered, in New York, Scotland, and Sweden (Kjellesvig-Waering, 1966). It was more than 50 years before another Silurian scorpion was recognized, and even this specimen had been collected in the late nineteenth century but misidentified as a eurypterid. Three species of scorpion were described from the Bertie Waterlimes of New York. The first was discovered in 1882 by OA Osborn but not reported in the literature (Whitfield, 1885a) until after the first description of one from Sweden. Whitfield placed the American scorpion into the same genus as the European one, calling it *Palaeophonus osborni* but, following this original notice Whitfield (1885b) gave a more detailed description and erected a new genus for the specimen *Proscorpius osborni*. Many more specimens of this species have since been collected (described in Kjellesvig-Waering, 1986) by Sam Ciurca. The second scorpion from the Bertie is *Archaeophonus eurypteroides*, described by Kjellesvig-Waering (1966), and the third, *Stoermeroscorpio delicatus*, was described by Kjellesvig-Waering in 1986, again, thanks to determined collecting by Sam Ciurca (**78**). They are known only from the Fiddlers Green Formation. All of these scorpions are rather primitive in form;

76 Giant chelicera of *Acutiramus cummingsi*. The specimen is a molt, as if it were a glove which has been pulled off. The hooked tips helped to catch prey and the huge teeth and saw-like tooth cut and sliced it up. The upper finger (ramus) of the claw is movable, and the long arm suggests that the claw was shot out rapidly to catch moving prey (YPM). Length of movable ramus 2.6 cm (1 in).

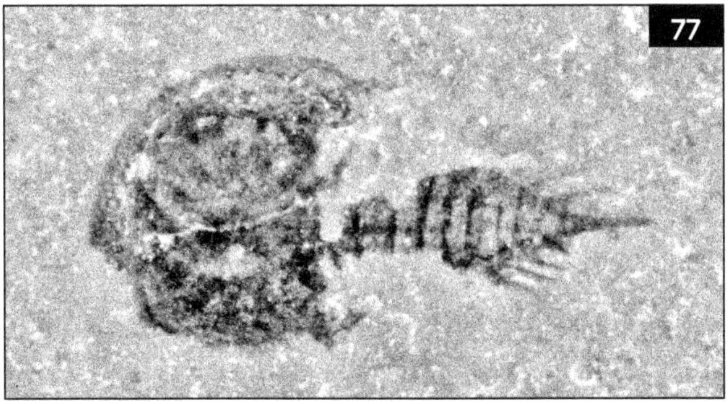

77 Horseshoe crab *Pseudoniscus clarkei*, Williamsville Formation bed A, Ridgemount Quarries, Ontario (YPM). Length of animal 16 mm (0.6 in).

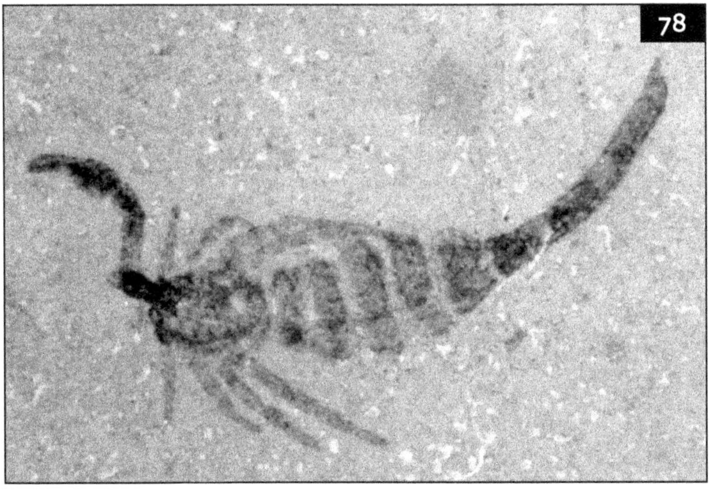

78 Scorpion *Stoermeroscorpio delicatus*, Phelps Waterlime Member, Fiddlers Green Formation, Passage Gulf, New York (YPM). Length of animal 12 mm (0.5 in).

they walked on the tips of their legs (digitigrade), bore gills, and lacked many terrestrial adaptations. So it was quite likely they were aquatic, like the eurypterids (Selden and Jeram, 1989). A new study of the Bertie scorpions by Dunlop *et al.* (2007) has shown that they all belong to one species: *Proscorpius osborni.*

Crustaceans

The phyllocarid *Ceratiocaris acuminata* (**79**) is common in the Williamsville Formation and somewhat rarer in the Fiddlers Green Formation of the Bertie Waterlime. These bivalved crustaceans are still around today, represented by only about 40 species in three families, but have a rich fossil record extending back to the Middle Cambrian Burgess Shale (Chapter 3). Characteristics of phyllocarids include the bivalved carapace which covers most of the body and appendages, and a telson (tail spine) usually with a furca attached.

Naraoia

First described from the Burgess Shale (Chapter 3), the strange arthropod *Naraoia*, with its uncalcified, two-part body over a trilobite-like underside, has generally been allied with the true trilobites. Since its first discovery in the Middle Cambrian Burgess Shale, *Naraoia*-like animals have been described from the Early–Middle Cambrian of Idaho and Utah, the Early Cambrian Emu Bay Shale of Australia, the Early Cambrian Chengjiang Biota of China, the Early Cambrian of Poland, the Early Cambrian of Sirius Passet, Greenland, the Early Ordovician of Sardinia, and the Late Ordovician Soom Shale of South Africa (see Selden and Nudds, 2004, Chapter 3). More recently, this record has been extended to the Silurian Period with the discovery of *Naraoia bertiensis* from the Williamsville Formation of the Ridgemount Quarries in Ontario (Caron *et al.*, 2004).

Budd (1999) described *Buenaspis* from the Lower Cambrian Sirius Passet fauna of Greenland, and also re-examined the naraoiid-like animals, concluding that they were better not included within the true trilobites (e.g. Fortey, 1997) but he resurrected an old taxon name, Nektaspida, for these animals. Caron *et al.* (2004) produced a phylogenetic analysis of all naraoiid-like animals and concurred with Budd. So, Nektaspida form a group of trilobite-like animals which are characterized by having no or very few (maximum five, in *Buenaspis*) free thoracic segments, an uncalcified dorsal exoskeleton, and no apparent eyes. The extension of the range by the find of a true *Naraoia* in the Bertie Waterlime shows that these near-trilobites were in existence far longer than previously thought and were most likely more diverse but rarely preserved owing to their soft exoskeleton.

Mollusks

Nautiloid cephalopods (**80**) are a characteristic element of the *Eurypterus lacustris* fauna of the Williamsville Waterlime. Both straight (orthocone) and coiled specimens are frequently discovered; some orthocones up to 1 m (3.3 ft) long have been encountered. High-spired gastropods are also common in the Williamsville; most are very small and usually occur in clusters.

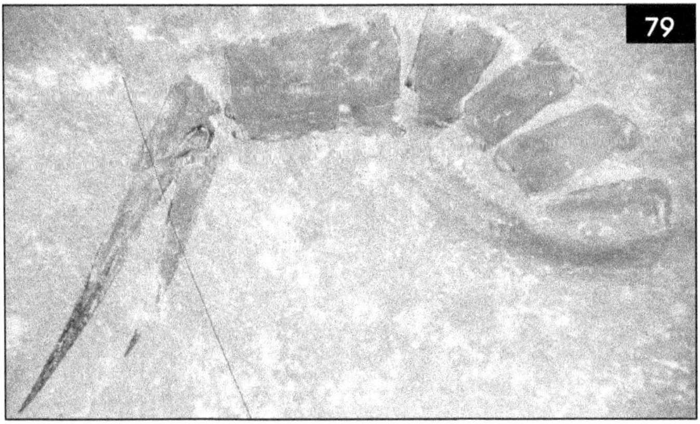

79 Tail of phyllocarid crustacean *Ceratiocaris acuminata*, Williamsville Formation, Ridgemount Quarries, Ontario (YPM). Length of tail-spike 6 cm (2.4 in).

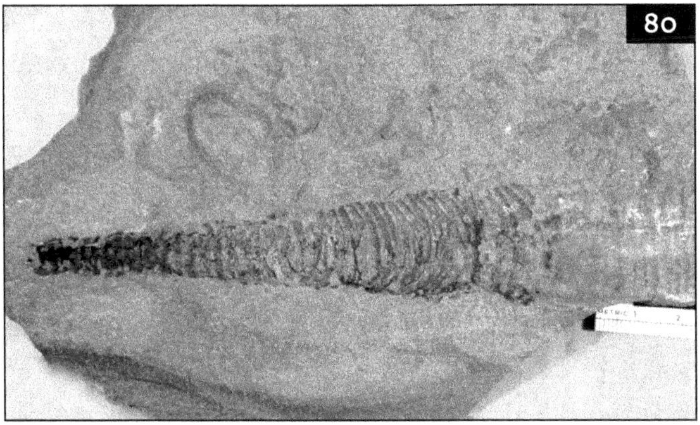

80 Orthocone nautiloid, Williamsville Formation bed A, Ridgemount Quarries, Ontario (YPM). Length 15 cm (6 in).

81 Lingulid brachiopods on and around *Eurypterus lacustris* carapace. Width of carapace 7.5 cm (3 in).

Brachiopods

The brachiopod *Whitfieldella* occurs throughout the Bertie Waterlime, but in many cases it occurs only as molds – the calcitic shell has been dissolved away, and in some beds this might have resulted in complete loss of shells. In others, it is abundant (see **67**). For example, in some layers of the Victor Member of the Fiddlers Green Formation *Whitfieldella* occurs as brachiopod 'pavements' – thin layers crowded with shells that are clearly not in life position. Such pavements occur when a storm erodes the sea floor and deposits the (now dead) shells as a single layer. The inarticulate brachiopod *Lingula* also occurs (**81**).

Other biota

A rather odd creature found occasionally in the Williamsville A Waterlime is called the 'sperm' by collectors (**82**), after its elongate shape with an apparent head and tail. It comes in various sizes, and one specimen shows a circular structure at the 'head' end and linear features along its length. A particularly large circular specimen (**83**), called 'Ezekiel's Wheel' by the collector, appears to be a giant form. We have no idea what this organism is nor, indeed, whether it was a free swimmer like a fish, attached to the sea floor like a coral, buried in the mud, or (and this is perhaps the most likely idea) it is part of a plant like *Cooksonia* (see below).

Land plants

Banks (1973) first mentioned the presence of the unequivocal land plant *Cooksonia* in the Bertie Waterlime. Until quite recently, the Bertie Waterlime was the only horizon yielding well-preserved Silurian plants with sporangia (spore-bearing organs) on the whole

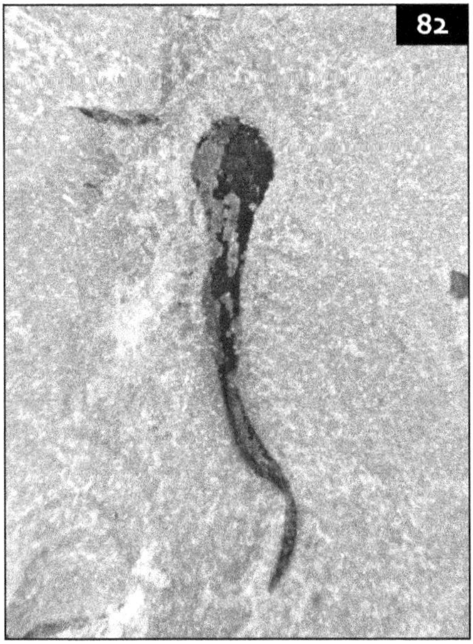

82 Small 'sperm', Williamsville Formation bed A, Ridgemount Quarries, Ontario (YPM). Length 3.5 cm (1.4 in).

83 'Ezekiel's Wheel', Williamsville Formation bed A, Ridgemount Quarries, Ontario (YPM). Length 4 cm (1.6 in).

palaeocontinent of Laurentia and is still the only locality in which to find Laurentian *Cooksonia*. Banks's original is a single specimen, without counterpart, and is preserved as an impression with patches of coalified residues on the dolomitic limestone. It consists of a cluster of axes (branches), three of which terminate in short, wide sporangia. Recently, many more specimens were discovered by Sam Ciurca, and were described by Edwards *et al.* (2004) (**84**).

Cooksonia is a simple vascular plant with smooth, presumably green stems which branch dichotomously (in equal parts, unlike most modern plants which have a main trunk and side branches). The clumps grow by branching and expand by creeping along the ground. Sporangia are borne at the tips of the fertile axes, rather than at the side, as occurs in higher plants.

84 The simple vascular plant *Cooksonia*, showing dichotomously branching axis and terminal sporangia. Williamsville Formation, Ridgemount Quarries, Ontario. Length 6.5 cm (2.6 in).

Stromatolites

A number of horizons in the Bertie Waterlime are characterized by hummocky, finely laminated structures known as stromatolites. One horizon particularly rich in stromatolite beds is the Ellicott Creek Breccia at the top of the Fiddlers Green Formation (**85**). Stromatolites are mats of lime mud precipitated by the action of photosynthesis by algae or cyanobacteria ('blue-green algae'). They can be found in rocks from as old as Precambrian to the present day because cyanobacteria and algae have been on Earth for such a long time (see Gunflint Chert, Chapter 1).

Paleoecology of the Bertie Waterlime

The Bertie Waterlime eurypterid horizons are fine-grained dolostones containing a mixture of clay and quartz and which break with a conchoidal fracture. Sedimentary features such as ripple marks, cross bedding, fine lamination, lunate scours, halite pseudomorphs, and trace fossils indicate that their deposition was in shallow water, between coastal stromatolite shoals and deep-water sponge reefs. The common halite psuedomorphs indicate that most – possibly all – waterlimes were formed under hypersaline conditions, and in cyclic sequences in an evaporating basin (Ciurca, 1973). The depositional environment is thus interpreted as a hypersaline lagoon. The question is: did the eurypterids live in this salty lagoon or were their remains washed in after death?

In other places, the presence of eurypterids is considered to indicate freshwater or brackish waters. Most (all?) of the eurypterids are molts, although Heubusch (1962) searched the collections of *Eurypterus lacustris* in the Buffalo Museum of Sciences and found two specimens, plus another in a private collection, which she considered showed evidence of a tubular structure (i.e. the gut) in the metasoma. Whether these specimens really represent dead animals is debatable, however, because: (a) arthropods molt all ectodermal structures, which includes the fore- and hind-gut; (b) judging from the photographs, the 'tubular structures' could just as easily be external carinae; and (c) dead animals are usually scavenged and so rarely turn up as fossils unless they have been quickly buried. Even if these three specimens do represent dead animals, the ratio of them to molts is clearly so small that we should consider that molts are the norm. While collectors favor complete specimens, most eurypterid remains in the rock are disarticulated parts that have been torn apart during transportation. Most eurypterid fragments occur in current-oriented debris accumulations known as windrows (**86**). Commonly they comprise the more resistant body parts such as telsons, and often there is taxonomic segregation so that, for example, pterygotids, cephalopods, or gastropods occur in discrete windrows.

Cooksonia was derived from the land, where it might have been growing in boggy, riverside habitats, yet it occurs together with aquatic animals in current-oriented deposits. So, many people consider that the eurypterids lived in adjacent rivers and their molted remains were washed in during occasional storms together with other nonmarine organisms. The hostile hypersaline conditions might have been conducive to the preservation of organic matter by decreasing the numbers of detritivores and bacteria. We should note, however, that the Bertie Waterlime sequence, as shown in (**67**) is complex, and many different micro-palaeoenvironments are represented, some of which contained good marine organisms (e.g. orthocone nautiloids) while others were washed in from elsewhere (e.g. *Cooksonia*).

85 Salt hopper crystal (next to 6 in (15 cm) rule) and stromatolitic mounds in Ellicott Creek Member (Fiddlers Green Formation), floor of Neid Road Quarry, Le Roy, New York.

86 Two *Eurypterus lacustris* specimens with a loose tergite, aligned in a 'windrow'. Largest specimen 17 cm (6.7 in) long.

Comparison of the Bertie Waterlime with other Lower Paleozoic biotas

Klussendorf (1994) compared the Bertie Waterlimes with other, similar Silurian biotas in the north-eastern states using cluster analysis and found that it grouped with other eurypterid-phyllocarid dominated localities such as Kokomo, Indiana. The *Eurypterus* Beds of the Silurian sequence in the Baltic region of Europe are remarkably similar in many ways to the Bertie Waterlimes. These are beautifully exposed on the Estonian island of Saaremaa and the Swedish island of Gotland. The beds in Gotland and Estonia appear to be of the same age (latest Wenlock) and differ in that the preservation of the eurypterids is rather better in the Gotland dolostones than those of Saaremaa (Selden, 1981). The Bertie Waterlimes are younger (Přídolí) than the Baltic sequence.

Today, because of the high level of plate tectonic activity which includes raised mid-ocean ridges, sea levels are higher than they were in the Silurian. The seas over the continental shelves now are deep, but during late Silurian times, vast, shallow seas occurred on the continental shelves, known as epicontinental seas. Both contain abundant specimens of *Eurypterus* and also pterygotid and other eurypterid species. In both cases the eurypterid remains are fragmentary and seem mostly to be molts. Both Bertie and Baltic dolostones also contain horseshoe crabs, lingulid and other brachiopods, orthocone nautiloids, gastropods and phyllocarid crustaceans. Both the Baltic region of Europe and New York lay together on the edge of the same continental land mass, Laurentia, so when we compare the two Lagerstätten, we are really comparing very similar ecosystems separated in space and time. The species of *Eurypterus*, *Pseudoniscus* etc. are different, just as *E. remipes* differs from *E. lacustris* because some morphological evolution occurred during the time between deposition of the Fiddlers Green and the Williamsville Waterlimes. The amount of evolution which occurred in the time between deposition of the Baltic beds and the Bertie was presumably rather more and so the species of *Eurypterus*, for example, are clearly separable.

Further Reading

Andrews, H. E., Brower, J. C., Gould, S. J. and Reyment, R. A. 1974. Growth and variation in *Eurypterus remipes* DeKay. *Bulletin of the Geological Institution of University of Uppsala: New Series 4* **6**, 81–114.

Banks, H. P. 1973. Occurrence of *Cooksonia*, the oldest vascular land plant macrofossil, in the Upper Silurian of New York State. *Journal of the Indian Botanical Society* **50A**, 227–235.

Braddy, S. J. and Dunlop, J. A. 1997. The functional morphology of mating in the Silurian eurypterid, *Baltoeurypterus tetragonophthalmus* (Fischer, 1839). *Zoological Journal of the Linnean Society* **121**, 435–461.

Budd, G. E. 1999. A nektaspid arthropod from the early Cambrian Sirius Passet fauna, with a description of retrodeformation based on functional morphology. *Palaeontology* **42**, 99–122.

Caron, J-B., Rudkin, D. M. and Milliken, S. 2004. A new late Silurian (Pridolian) naraoiid (Euarthropoda: Nektaspida) from the Bertie Formation of southern Ontario, Canada – delayed fallout from the Cambrian Explosion. *Journal of Paleontology* **78**, 1138–1145.

Chapman, E. J. 1864. *A Popular and Practical Exposition of the Minerals and Geology of Canada.* Toronto.

Ciurca, S. J. 1973. Eurypterid horizons and the stratigraphy of Upper Silurian and Lower Devonian rocks of western New York State. *New York State Geological Association 45th Annual Meeting Field Trip Guidebook* D1–D14.

Ciurca, S. J. 1990. Eurypterid biofacies of the Silurian – Devonian evaporite sequence: Niagara Peninsula, Ontario, Canada and New York. *New York State Geological Association, 62nd Annual Meeting Field Trip Guidebook* D1–D30.

Ciurca, S. J. and Hamell, R. D. 1994. Late Silurian sedimentation, sedimentary structures, and paleoenvironmental settings within an eurypterid-bearing sequence (Salina and Bertie Groups), western New York State and southwestern Ontario, Canada. *New York Geological Association 66th Annual Meeting Field Trip Guidebook* 455–488.

Clarke, J. M. 1902. Notes on Paleozoic crustaceans. *New York State Museum Report* (1900) **54**, Appendix 3, 83–110.

Clarke, J. M. 1919. *Bunaia woodwardi*, a new merostome from the Silurian waterlimes of New York. *Geological Magazine, decade 6* **6**, 531–532.

Clarke, J. M. and Ruedemann, R. 1912. The Eurypterida of New York. *New York State Museum Memoir* **14**.

Cuggy, M. B. 1994. Ontogenetic variation in Silurian eurypterids from Ontario and New York State. *Canadian Journal of Earth Sciences* **31**, 728–732.

Dunlop, J. A., Tetlie, O. E. & Prendini, L. 2007. A reinterpretation of the Silurian scorpion *Proscosrpius osborni* (Whitfield, 1885): integrating data from Paleozoic and Recent scorpions. *Palaeontology*. In press.

Edwards, D., Banks, H. P., Ciurca, S. J. and Laub, R. S. 2004. New Silurian cooksonias from dolostones of north-eastern North America. *Botanical Journal of the Linnean Society* **146,** 399–413.

Eldredge, N. 1974. Revision of the suborder Synziphosurina (Chelicerata, Mersotomata), with remarks on merostome phylogeny. *American Museum Novitates* **2543,** 1–41.

Fortey, R. A. 1997. Classification. 289–302. In: *Treatise on Invertebrate Paleontology*. Part O. Arthropoda 1, Trilobita (revised). R. L. Kaesler (ed.). Geological Society of America and University of Kansas, Boulder, Colorado and Lawrence, Kansas.

Heubusch, C. A. 1962. Preservation of the intestine in three specimens of *Eurypterus*. *Journal of Paleontology* **36,** 222–224.

Kjellesvig-Waering, E. N. 1958. The genera, species and subspecies of the family Eurypteridae, Burmeister, 1845. *Journal of Paleontology* **32,** 1107–1148.

Kjellesvig-Waering, E. N. 1966. Silurian scorpions of New York. *Journal of Paleontology* **40**, 359–375.

Kjellesvig-Waering, E. N. 1986. A restudy of the fossil Scorpionida of the world. *Palaeontographica Americana* **55**, 1–287.

Kjellesvig-Waering, E. N. and Heubusch, C. A. 1962. Some Eurypterida from the Ordovican and Silurian of New York. *Journal of Paleontology* **36**, 211–221.

Kluessendorf, J. and Mikulic, D. G. 1991. The role of anoxia in the formation of Silurian Konservat Lagerstätten. *Geological Society of America, Northeastern and Souteastern Sections, Baltimore, United States, Abstracts with Programs* **23**, 54.

Kluessendorf, J. 1994. Predictability of Silurian Fossil-Konservat-Lagerstätten in North America. *Lethaia* **27,** 337–344.
Milne, L. and Milne, M. 1965. *The Crab that Crawled out of the Past.* Atheneum, New York.
Mitchill, S. L. 1818. An account of the impressions of a fish in the rocks of Oneida county, New York. *American Monthly Magazine* **3**, 291.
Nieszkowski, J. 1858. *De Euryptero remipedo.* Dorpat.
Nieszkowski, J. 1859. Der *Eurypterus remipes* aus den obersilurischen Schichten der Insel Oesel. *Archiv für Naturkunde der Liv-, Est- und Kurlands, Series 1* **2**, 299–344. 2 pls.
Rickard, L. V. 1975. Correlation of the Silurian and Devonian rocks in New York State. *New York State Museum Map and Chart Series* **24**, 1–16.
Ruedemann, R. 1916. Account of some new or little-known species of fossils, mostly from Paleozoic rocks of New York. *New York State Museum Bulletin* **189**, 7–112.
Ruedemann, R. 1925. Some Silurian (Ontarian) Faunas from New York State. *New York State Museum Bulletin* **265**, 1–134.
Selden, P. A. 1981. Functional morphology of the prosoma of *Baltoeurypterus tetragonophthalmus* (Fischer) (Chelicerata: Eurypterida). *Transactions of the Royal Society of Edinburgh: Earth Sciences* **72**, 9–48.
Selden, P. A. and Jeram, A. J. 1989. Palaeophysiology of terrestrialization in the Chelicerata. *Transactions of the Royal Society of Edinburgh: Earth Sciences* **80**, 303–310.
Selden, P. and Nudds, J. 2004. *Evolution of Fossil Ecosystems.* Manson, London.
Størmer, L. and Kjellesvig-Waering, E. N. 1969. Sexual dimorphism in eurypterids. In: *Sexual Dimorphism in Fossil Metazoa and Taxonomic Implications.* G. E. G. Westermann (ed.). International Union of Geological Sciences, Series A **1**, 201–214.
Tetlie, O. E. 2006. Two new Silurian species of Eurypterus (Chelicerata: Eurypterida), from Norway and Canada and the phylogeny of the genus. *Journal of Systematic Palaeontology* **4**, 397–412.
Tetlie, O. E., Tollerton, V. P. and Ciurca, S. J. 2007. The Silurian eurypterids (Chelicerata: Eurypterida) *Eurypterus remipes* and *E. lacustris* from New York State, USA and Ontario, Canada, are separate species. *Bulletin of the Peabody Museum of Natural History,* **48**, 139–152.
Tollerton, V. P. 1992. Heterochrony in eurypterids and the ontogenetic resolution of morphological differences in *Eurypterus remipes* and *E. lacustris. Geological Society of America, Abstracts with Programs* **34**, 98.
Whitfield, R. P. 1885a. An American fossil scorpion. *Science* **6,** 87.
Whitfield, R. P. 1885b. On a fossil scorpion from the Silurian rocks of America. *Bulletin of the American Museum of Natural History* **1**, 181–190.

Gilboa

Chapter 6

BACKGROUND: COLONIZATION OF THE LAND

The consequences of plants and animals leaving the marine realm to colonize the land were far-reaching, not least for the evolution of Man, who is part of the Earth's terrestrial biota. Few of the metazoan phyla which emerged from the great Cambrian explosion produced terrestrial forms, but those which did have became very successful in terms of diversity. The Arthropoda includes some terrestrial crustaceans (e.g. woodlice), but of much greater importance are the terrestrial chelicerates (spiders, scorpions, mites, and their allies) and the insects, which make up 70% of all animals alive today. From fish arose the tetrapods: amphibians, reptiles, birds and mammals. Mollusks, too, in the form of slugs and snails, have been remarkably successful on land, as any gardener will testify. In order to live successfully on land, plants developed features such as stiff trunks and reproductive devices which gave rise to the familiar trees and flowers we see on land today.

Terrestrialization was thus a major episode in the evolution of life on Earth. It was neither an instantaneous event nor restricted to a particular geological period; indeed some organisms, such as crabs, may be considered to be terrestrializing today. Nevertheless, when the physical conditions on the land surface became sufficiently favorable to life in the Silurian Period, then the invasion began in earnest. A number of physical barriers have to be overcome before an organism can make the jump from life in the sea to land. Water is necessary for all biological processes but its supply is variable on land compared to the sea. Plants and animals adopt four main strategies to cope with under- (or over-) supply of water. Some, like microbes, ostracodes, and algae, live in permanent water on land, between soil particles and in ponds, so they are effectively aquatic. Others, like amphibians, millipedes, slugs, and woodlice, live in moist habitats and only venture out into dry air for short periods. Poikilohydric organisms can tolerate desiccation and rehydrate when necessary; examples are bryophytes (mosses and liverworts), and 'resting' stages such as plant spores, fairy shrimp eggs, and tardigrade ('water-bear') tuns. This group was probably the first onto land in the Paleozoic Era. The most successful terrestrial organisms of all are the homoiohydric forms, which maintain permanent internal hydration mainly by having a waterproof cuticle or skin. These are the familiar land plants (tracheophytes), tetrapods, and the arthropods.

Another physiological barrier to be overcome when crossing the threshold from water to land is breathing or, more specifically, exchange of the gases oxygen O_2, and carbon dioxide CO_2. Both plants and animals exchange O_2 and CO_2 with the environment across a semi-permeable membrane but, because the O_2 and CO_2 molecules are both larger than the water molecule, H_2O, this membrane will leak, resulting in a loss of water. To overcome this problem, land plants have small holes, stomata, in their waterproof cuticle, which can be closed to prevent excessive water loss. Animals, which might have used external gills when living in water, have lungs or tracheal systems (in insects, spiders, and millipedes, for example) for breathing air enclosed within the body, and connected to the air by small holes (called spiracles in insects).

Other adaptations which evolved during terrestrialization include: stronger legs and better balance to account for the loss of buoyancy; sense organs which operate in a medium with different optical and acoustic properties (sound is used more for communication on land); more careful ionic balance, which is linked with the reduced availability of water on land; and the development of direct copulation during mating (in water, gametes can simply be discharged without the sexes coming into physical contact). In spite of all these problems, organisms swarmed onto the land. It was, after all, an unexploited ecological niche and there was, at least at first, some escape from predators in the sea.

History of discovery of Gilboa

The first settlement on the banks of Schoharie Creek where it rushes through gorges between the junctions of the Manorkill and Plattekill tributaries was established in 1764 by Matthew and Jacob Dies. Where a river flows quickly, in a gorge or over a waterfall, is an ideal spot for the siting of a mill. As the number of mills grew, the town of Gilboa was established, presumably named after the biblical Mt Gilboa in Palestine (Hernick, 1996: this book gives a more detailed and authoritative history of Gilboa fossil studies). Fossil plant material is abundant in the Devonian rocks of the area, and the first district to be surveyed in the geological survey of New York was the Catskill Mountains (**87**). The publication arising from the survey, by Mather (1843), was the first to describe and

87

87 The Catskill Mountains: a fall view from near South Mountain Quarry.

figure fossil stems and leaves from outcrops in the Manorkill and Schoharie Creeks near Gilboa (**88**).

In 1852, Samuel Lockwood came to Gilboa as minister of the Reformed Church. Lockwood stayed for less than two years, but during that time his interest in geology and natural history led him to walk through the countryside and examine the rock outcrops, where he found the natural cast of a tree stump and other plant fossils. One of his specimens was eventually described by the famous Canadian paleobotanist John W Dawson as *Caulopteris lockwoodi* (Dawson, 1871). In October 1869, one of the regular floods which sweep down Schoharie Creek devastated the riverside mills, roads, and bridges. A gang of men working on the bridge repairs, blasting the rock outcrop to establish footings, uncovered a veritable forest of tree stumps in growth position. The find came to the attention of James Hall, then State Paleontologist, who immediately organized to bring several stumps to the State Museum of Natural History in Albany. Dawson (1871) described the stumps as belonging to the tree fern *Psaronius*. A new tree stump locality, at Manorkill Falls, was discovered in 1895 by Charles S Prosser, during a new state geological survey of the Devonian rocks in eastern New York.

As the century turned, little more geological work was done in the area, but the growing city of New York looked to the Catskills as a source of water for its ever-expanding population. A New York

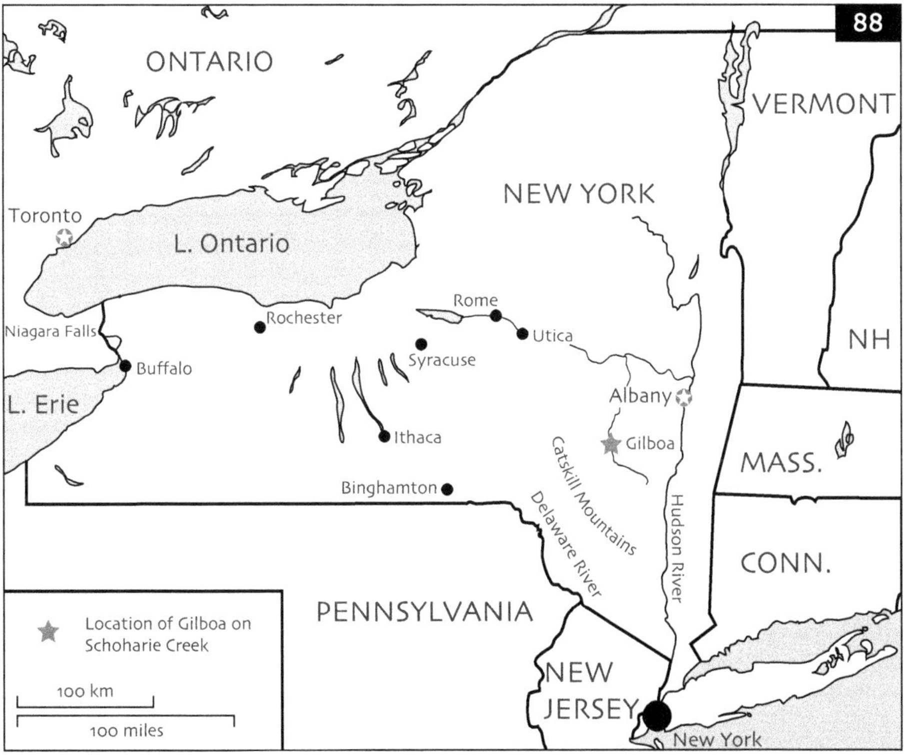

88 Map of New York, showing the location of Gilboa.

State Legislature act of 1905 enabled the acquisition of lands and building of dams and aqueducts for the purpose of supplying water to the city. In 1917, construction started on a dam across Schoharie Creek at the Gilboa gorge. The reservoir drowned the former settlement of Gilboa, and 935 bodies had to be reinterred from its cemetery (Hernick, 1996). The water supply for agriculture, the fishing, and the tourist trade all suffered as river levels dropped.

Paradoxically, it was the threat of loss of the Gilboa fossil forests under the new reservoir that provoked the New York State Geological Survey in 1920 into conducting a field survey to locate and rescue any important plant fossils. An outstanding discovery of fossil stumps was made at the lower Manorkill Falls, where Prosser had found them in 1895, and Rudolf Ruedemann discovered some fossil 'seeds' (actually, sporangia, see later) at the falls during the same trip. Ruedemann (1926, pp. 510–511) gave an interesting account of the finding of the 'seeds': "One night, as we were trying our luck at fishing in the Schoharie Creek, both my companions went out on a bowlder in the river while I remained on shore to enjoy the beautiful scenery of the Manorkill Falls just behind us. While I was sitting there, I noticed a large slab of black shale sticking out of the river sand. It was covered with beautiful clusters of fern seeds. I seized it, and yelled to the fishermen. We carried the precious slab home. When we arrived, my companions discovered they had left their pipes and tobacco on the bowlder and blamed me for having made too much 'fuss' over my find. But it was worth it, for these are the oldest seeds known at present. There was no doubt that the bed from which the slab came was near by, and we found it the next morning not a hundred feet away at the foot of the cliff in the corner between the Manorkill and Schoharie creeks. From it we secured a fine collection of seeds. Miss Goldring went out later and obtained the spore-bearing organs, the foliage and rootlets. The locality will be lost to science when the reservoir is filled."

The 'Miss Goldring' referred to by Ruedemann was Winifred Goldring who was responsible for an important contribution to the paleontology of New York through the first half of the 20th century, eventually becoming State Paleontologist after John M Clarke. Goldring had a good knowledge of botany, so the Gilboa stumps were an ideal project for her to embark upon, after finishing her major monograph on the Devonian crinoids of New York. She was responsible for the display of the Gilboa fossil forest in the New York State Museum in Albany. Goldring had much more plant material on which to work than did Dawson. She hypothesized that the various parts (leaves, seeds, stumps) belonged to the same plant, which she surmised was a seed-fern (pteridosperm), not a tree fern. She produced a reconstruction of the plant as a tree-sized seed-fern, and called it *Eospermatopteris*. In addition to the fossil forest exhibit in Albany, Miss Goldring arranged for some stumps to be placed as a public exhibit by the road outside Riverside Quarry, which had produced the greatest number (and biggest at 3.3 m [11 ft] in circumference) of stumps. The quarry has long since been filled in, but the Gilboa Historical Society has relocated the original stumps to near the town hall and created a new exhibit (**89**).

German paleobotanists Richard Kräusel and Herman Weyland went to the United States and Canada to study Devonian plant fossils, including the Gilboa *Eospermatopteris*. These authors compared Gilboa *Eospermatopteris* to similar plants from Germany. They concluded that the supposed 'seeds' were actually spore-bearing organs (sporangia) and the plant was not a seed-fern but belonged to a group known as progymnosperms, similar to the German plant that Kräusel and Weyland (1923) called *Aneurophyton*. More recently, Serlin and Banks (1978) looked again at *Aneurophyton*, specimens of which they

89 Stumps of *Eospermatopteris* from Riverside Quarry collected by Miss Goldring now forming an exhibit outside the town hall at Gilboa.

had collected from a disused quarry on the New York bank of the Delaware. They concluded that the foliage and sporangia of the New York material were indeed *Aneurophyton*, but the foliage could not be linked directly with the stumps. Though commonly found with the stumps, the association could be ecological rather than anatomical. In paleobotany, different parts of the plant receive different names when they are described separately; it is not until somebody proves that were attached in life that one name can be subsumed under the other (the earliest name takes priority). The mystery was solved by Stein *et al.* (2007) who described a spectacular specimen of *Eospermatopteris* with cladoxylopsid fronds (see later).

Through the middle and later parts of the twentieth century, there was an active group of paleobotanists researching the fossil flora of the Devonian rocks of new York, headed by Harlan P Banks of Cornell University. A monograph on the lycopods (club-mosses) of New York was produced by Banks and his student James D Grierson (Grierson and Banks, 1963). Riverside Quarry was a particularly good locality for club-mosses. The site has yielded two genera of herbaceous club-mosses (*Gilboaphyton* and *Protolepidodendron*) and three arborescent forms (*Amphidoxodendron*, *Sigillaria*, and *Lepidosigillaria*), as well as the progymnosperm *Aneurophyton* and four genera belonging to enigmatic plant groups (*Eospermatopteris*, *Pseudosporochnus*, *Ibyka*, and *Prosseria*) (Berry and Fairon-Demaret, 2001). These are described later.

Another locality which has yielded not only fossil plants, but animals as well, is known as Brown Mountain. As New York city grew, its appetite not only for water but also for electric power grew with it. In 1968 the New York Power Authority was given permission to create a pumped-storage power scheme on Brown Mountain, just north of Gilboa near the township of North Blenheim. New works were constructed (**90**), so that water can be pumped to a holding reservoir at the top of the mountain using spare power capacity in slack times, which can then be released through turbines back down to the Blenheim–Gilboa Lower Reservoir when the grid needs additional power. Yet again, the residents of Gilboa suffered the loss of farmland and some homes, in addition to disruption during the building phase. However, the Power Authority was very helpful to paleontologists, and there is now a visitor center which explains the electricity project from a vantage point overlooking the reservoirs, and also gives some information on the fossils discovered during the works. During construction of the dam for the lower reservoir, a lens of rock packed with plant stems was discovered by local schoolteacher Raymond A. Baschnagel. The find was studied by Banks, Grierson, and colleague Patricia M. Bonamo from the State University of New York at Binghamton (Banks *et al.*, 1972), who erected a new genus and species, *Leclercqia complexa,* for the plant stems. What was significant about their work was that they were able to macerate the rock, a fine mudstone, in hydrofluoric acid (HF), which dissolves the siliceous matrix but leaves the organic matter intact. Normally, with burial under thousands of feet of overburden, organic matter loses its volatiles (as oil and gas) and reduces to carbon. At the Brown Mountain locality, the organic matter is still a brown colour, rather than black carbon, and so fine details of the plant anatomy could be seen under the microscope. *Leclercqia* is an herbaceous lycopod; other plants found at this locality were *Haskinsia,* another lycopod, and *Rellimia,* a progymnosperm.

The most exciting finds at Brown Mountain, however, were excellently preserved remains of tiny arthropods: various arachnids such as scorpions, pseudoscorpions, mites, trigonotarbids, and spiders; centipedes and other myriapods; and possibly the earliest insect relatives. At that time (1970s) the oldest known land animals came from the early Devonian site at Rhynie, Scotland (Selden and Nudds, 2004, Chapter 5); although younger than these, the Gilboa material represented the earliest known land animals ever found in North America (Shear *et al.*, 1984). The preservation of the animals was so remarkable that Bonamo and Grierson had to make sure that they were not looking at contamination by modern animals which had fallen into the acid during

90 View of the lower Schoharie Reservoir and powerhouse at Brown Mountain from the Blenheim – Gilboa Visitor Centre.

preparation. They took slides bearing the animal pieces to various experts on fossil arthropods, and the importance of the find was immediately recognized by W. D. Ian Rolfe, then of the Hunterian Museum, Glasgow, Scotland. Rolfe realized that, with so many different animal taxa present, he needed expert advice from workers on mites, other arachnids, and myriapods to do justice to this material. He showed pictures to William A. Shear, of Hampden-Sydney College, Virginia, an expert not only on modern arachnids but on myriapods as well. Shear and Rolfe became excited about the find, reported it in the journal *Science* with Bonamo, Grierson, and other colleagues (Shear *et al.* 1984), and then set to work describing the fauna group by group.

Following the discovery of the exceptional early terrestrial fauna at Brown Mountain, a number of other sites in the Catskills were searched for similar animal remains. So far, one other site has proved fruitful: a shale quarry in slightly younger beds near Conesville, east of Gilboa, referred to as South Mountain. The Gilboa area localities are of international importance because there are so few places in the world where a nearly complete early terrestrial ecosystem has been preserved and can be studied. Gilboa gives a rare insight into what the life on land was like shortly after it first emerged from the sea. Important though the Brown Mountain site is, it is now buried under tons of concrete of the Blenheim–Gilboa Reservoir. Bonamo and Grierson rescued as much of the lens of rock as they could, which is now stored at Binghamton, and where it was worked on for several years after the loss of the site.

Stratigraphic setting and taphonomy of Gilboa

It will be clear from the foregoing that we are actually dealing with a number of localities in the area around Gilboa, at least three of which are particularly important: Riverside Quarry, Brown Mountain, and South Mountain. As mentioned in the last section, the preservation of fossils at the Brown Mountain locality is rather unusual in that the organic matter is not oxidized or carbonized but retains its complexity. It is not the original plant and animal cuticles that are preserved, however, but randomly polymerized complex organic molecules. Hence, the fragments retain a brown coloration, some transparency and flexibility, despite being flattened in the fine mudstone. This means they show incredible details under the light microscope but are not three-dimensional except in occasional cases where iron pyrite has infilled spaces and cells. For paleobotanists, the variety of preservational types, combined with a range of techniques – light microscopy, scanning electron microscopy (SEM), and chemical preparation techniques – provided them with exceptional views of early land flora. For the paleozoologists, light microscopy, rather than SEM, proved to be the ideal method to view the fauna.

At Riverside Quarry and other localities that yield stumps (Goldring, 1924, 1927; Driese *et al.*, 1997) the plant foliage is carbonized, but the stumps are sandstone casts. The stumps are preserved in life position, seated on a paleosol (fossil soil) represented by a 15–60 cm (6–24 in) thick claystone, and surrounded by coarse, cross-bedded sandstone. Adjacent to the stumps, Goldring found loose trunks up to 4 m (>12 ft) in length. In one small area, 18 stumps were found, 1–2 m (3–6.5 ft) apart. The study by Driese *et al.* (1997) showed that the roots radiating up to 3 m (9 ft) from the stumps into the paleosol were shallow, and the gley nature of the paleosol indicated waterlogged conditions. We can envisage a swamp forest of *Eospermatopteris* trees, similar to the Swamp Cypress (*Taxodium*) forests of the south-eastern seaboard states today, which became inundated by breaches of the levee or coastal barrier, bringing in cross-bedded sands. The inundation eventually killed the trees, their trunks gradually died and snapped off, then further influxes of sand filled the empty moulds left by the decayed tree stumps.

The Brown Mountain strata belong to the Panther Mountain Formation of the Hamilton Group, of Givetian, upper Middle Devonian, in age (**91**) (Sevon and Woodrow, 1985). The Riverside Quarry and Manorkill Falls fossil forests belong to the slightly younger Moscow Formation of the Hamilton Group (Banks *et al.*, 1985). The still younger deposit at South Mountain, belonging to the Oneonta Formation, Genesee Group (latest Givetian–earliest Frasnian) has also yielded fossil fauna (Shear and Selden, 1995, 2001) as well as abundant flora (Hueber, 1960; Carluccio *et al.*, 1966; Hueber and Banks, 1979). In addition, there are many, many small quarries and roadcuts around the Catskills, mainly for shale to mend the dirt roads in this area, which yields fossil plants.

The Hamilton Group sediments in eastern New York consist of sandstones and shales derived from erosion of a chain of mountains to the east. The mountains were rising and, simultaneously, being eroded, due to the action of the Acadian Orogeny. In Chapter 5, **67**, the top of the Silurian sequence shows a marked unconformity, with the oldest Devonian beds being laid down a long time after the end of the Silurian Period. At the end of the Silurian Period, the microcontinent Avalonia collided with the edge of the continent Laurasia on which the present-day New York lay. As subduction continued, so a chain of mountains was created parallel to the continental edge. The situation would have resembled that which we see today where the Indian tectonic plate continues to attempt to subduct beneath the Himalaya mountain belt, which is still rising but also being deeply eroded at the same time, producing a mass of sediment which is pouring down into the Indian Ocean, forming the mighty Ganges and Brahmaputra river deltas. Similarly, as the Acadian mountain chain was eroded, so rivers draining to the west poured their sediments into the marine basins of New York, building up the vast Catskill delta. Uplift periods were episodic, resulting in cycles of shallow-marine limestones (e.g. the Onondaga limestones, **91**), followed by deeper marine shales (e.g. the Marcellus shales), then coarse clastics (e.g. the Hamilton Group) as a delta lobe pushed westward into the marine basin. Each cycle was then interrupted by uplift (unconformity) and then repeated (e.g. Tully Limestones).

Description of the Gilboa biota

Plants

Zosterophyllopsida

These are the most primitive vascular plants to be found in the New York Devonian, and are characterized by lateral sporangia (i.e. sporangia borne on the sides of the stem (axis) rather than the tip, as in even more primitive plants), and an elliptical vascular strand (the cross-sectional shape of the water-conducting tissue in the axis). By Middle Devonian times, such forms were beginning to become extinct as they were outcompeted by more modern groups. The best known zosterophyll in the New York Devonian is *Serrulacaulis* (Hueber and Banks, 1979), which occurs at South Mountain. Specimens of *Sawdonia* (formerly called *Psilophyton*) have been found at the same locality, but only sterile axes, i.e. lacking the characteristic sporangia (Hueber and Grierson, 1961).

Lycopsida

The commonest group of plants in the Gilboa beds are the club-mosses or lycopods (Lycopsida); together with zosterophylls, they form the sister-group to most other land plants. At the present day, club-mosses are small, creeping forms with erect fertile branches maximally 30–60 cm (1–2 ft) high, e.g. the ground pines (*Lycopodium*, *Selaginella*) and ground cedars (*Diphasiastrum*, **92**). Many of the Gilboa lycopods were no bigger, but some reached tree-sized proportions, and by the Pennsylvanian Period they were the

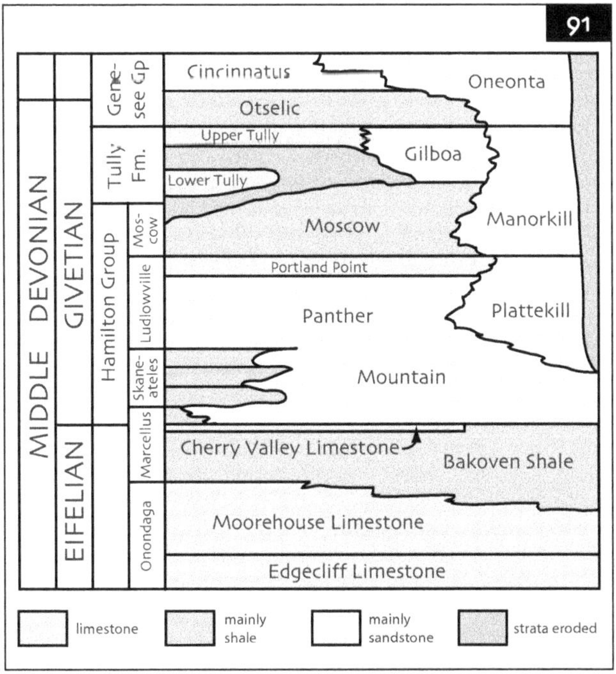

91 Stratigraphy of the Middle Devonian of the Gilboa area, New York (after Sevon and Woodrow, 1985).

92 A modern lycopod, ground cedar (*Diphasiastrum digitatum*), growing in Virginia. Height approximately 14 cm (5.5 in).

major trees in the coal forests. Arborescent (tree-like) lycopods at Gilboa include *Amphidoxodendron*, ?*Sigillaria*, and *Lepidosigillaria* (Grierson and Banks, 1963). *Lepidosigillaria* is one of the best-known Devonian trees, and a famous example was found at the town of Naples, in the Finger Lakes region of western New York, in black mudstone (Banks, 1966). The trunk, some 4 m (12 ft) long, is preserved in the New York State Museum. Presumably, this trunk floated out from the Catskill delta until it became waterlogged and sank into the deep sea to be covered by dark mud.

Herbaceous lycopods include *Gilboaphyton goldringiae*, described by Arnold (1937), *Protolepidodendron gilboense*, a Grierson and Banks (1963) species, *Haskinsia* and *Leclercqia complexa*, the anatomy of which was uncovered by the painstaking work of Banks, Bonamo, and Grierson referred to above (Banks *et al.*, 1972). *Gilboaphyton*, the leaves of which have two lateral teeth, has since been found in the Devonian of Venezuela (Berry and Edwards, 1997). The spores of *Leclerqia* were studied by Richardson *et al.* (1993). Spores can be important for establishing stratigraphy where other zonal indicators (e.g. marine invertebrates) are absent. However, study of spores within sporangia are necessary to identify correctly to which plant the abundant dispersed spores in the rock belong. *Leclercqia* has hook-like, recurved leaves with four lateral branches and it is this characteristic which produced the tangled masses of the plant, found in the rocks at Brown Mountain, which acted as a mesh to entrap the animal remains which make the locality an important Lagerstätte for Devonian land animals.

Progymnosperms

These are the most advanced trees found in the New York localities described here. As the name suggests, their morphology is part-way between the spore-bearing plants and the more advanced seed-bearing gymnosperms: progymnosperms bear sporangia and have leafy branches on large trunks whose wood is gymnospermous in nature. The foliage of *Aneurophyton*, originally described by Goldring (1924) from the stump localities, is progymnospermous, as is *Rellimia*, another common progymnosperm from Brown Mountain. Their axes show more complex branching than the simple dichotomous patterns seen in more primitive vascular plants. Both of these genera may well have been herbaceous rather than trees. The well-known progymnosperm *Archaeopteris* occurs in abundance at South Mountain (**93**). It was an arborescent progymnosperm with leafy branches on large trunks similar to modern trees.

Cladoxylopsida

Details of the internal anatomy of this group of Devonian plants are confusing regarding their placement among other plant groups. They show the most complex vascular strand patterns of all the Devonian plant groups, and are also the commonest of the arborescent forms.

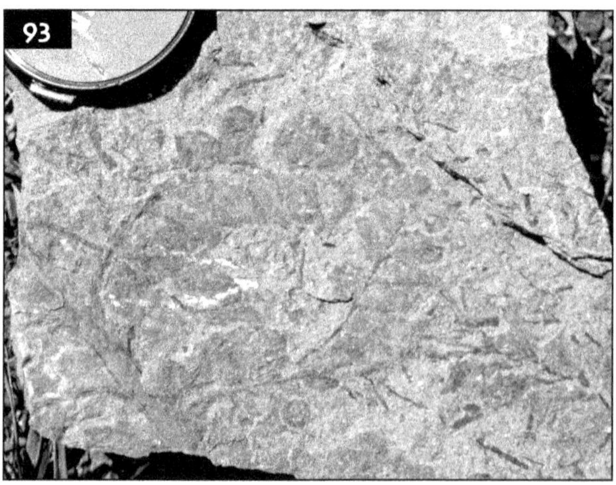

93 Foliage of the progymnosperm *Archaeopteris* at South Mountain. Lens cap diameter 6 cm (2.4 in).

Pseudosporochnus is the best known cladoxylopsid foliage, and this genus has been found at Riverside Quarry. The large stumps described by Goldring (1924) as *Eospermatopteris* are also cladoxylopsid.

Other plants

Ibyka, a plant probably intermediate between the lycopsids and progymnosperms, possibly ancestors of horsetails (Skog and Banks, 1973; **94**) and *Prosseria* (Banks, 1966) are both known from Riverside Quarry (Berry and Fairon-Demaret, 2001).

Animals

Eurypterids

A few patches of cuticle and commoner fragments with a striking resemblance to the gill tracts of *Eurypterus tetragonophthalmus*, described by Wills (1965), have been found in the macerate from Brown Mountain, as well as some characteristic leg segments (podomeres, **95**).

94 The modern horsetail *Equisetum* growing in a damp situation. Height about 60 cm (2 ft).

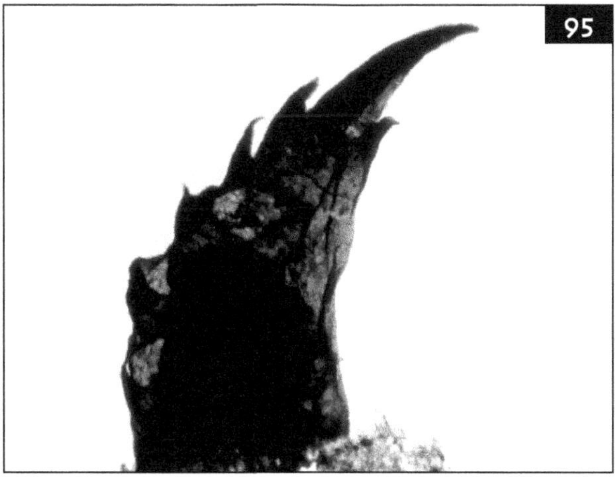

95 Possible tip of eurypterid leg (AMNH). Length 0.65 mm (0.03 in).

Scorpions

Some pieces of body segments and podomeres attributable to scorpion have been found at Brown Mountain and some palpal chelae (claws) and pectines have been found at South Mountain, though none of these has yet been formally described.

Trigonotarbids

These animals are very closely related to spiders but differ from them in lacking poison and silk glands, and are rather more primitive in other ways. Trigonotarbids were already known to be among the earliest land animals since they were described from the early Devonian Rhynie Chert of Scotland (Selden and Nudds, 2004, Chapter 5), but their discovery in the mudstones of Brown Mountain, Gilboa (Shear *et al.*, 1984) was first evidence of such early land life in North America. The Rhynie trigonotarbids are preserved three-dimensionally in translucent chert, and studies had already revealed their two pairs of book-lungs – enclosed lamellate organs for breathing air – which proved they were terrestrial. The Gilboa material is extremely flattened (**96, 97**), but looking at slides of this remarkably well-preserved material you get an eerie impression of how the

96

96 Nearly complete trigonotarbid body (legs missing) *Gilboarachne griersoni*; these specimens are completely flattened but under the light microscope at high magnification fine details can be seen (AMNH). Length 2.3 mm (0.09 in).

97
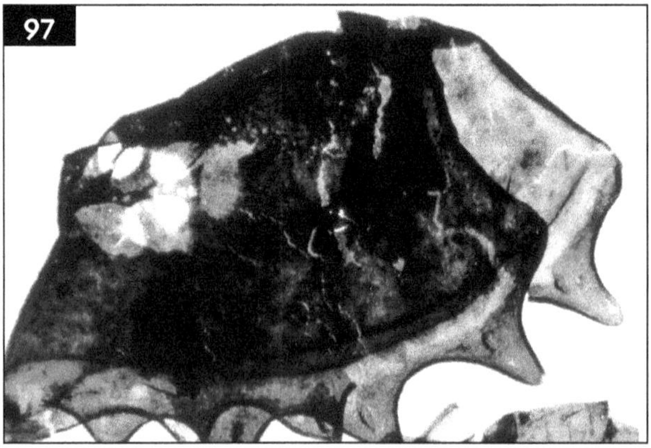

97 Carapace of the trigonotarbid *Gelasinotarbus reticulatus*, flattened laterally but giving the appearance of its three-dimensional shape in life. Note the pair of compound eyes (left), each with three major lenses and minor lenses between, and deeply scalloped carapace margin (AMNH). Length 3.1 mm (0.1 in).

animals might have looked in life (**98**). Trigonotarbids have a tough carapace with projections between the leg bases. There are two clusters of eyes, each consisting of three major lenses and several minor ones between, and a pair of median ocelli. The eye clusters show an evolutionary stage between compound eyes with many, similar-sized lenses, and the condition in spiders (which have three major eyes in each lateral group plus the median ocelli = eight altogether). The first pair of appendages is the chelicerae: little pincers, as found in all chelicerates (see Chapter 5); all other appendages on the prosoma are leg-like. The opisthosoma, or abdomen, is covered with hard plates, presumably for protection from predators. Several genera and species of trigonotarbid were described from Brown Mountain (Shear *et al.*, 1987), and others have been found elsewhere in the Devonian of the Catskills, e.g. South Mountain, and in the late Devonian of Pennsylvania (Shear, 2000).

Spiders

Shortly after the description of the trigonotarbids from Brown Mountain, a fragment of cuticle was discovered which bore silk-producing spigots (**99**), and thus

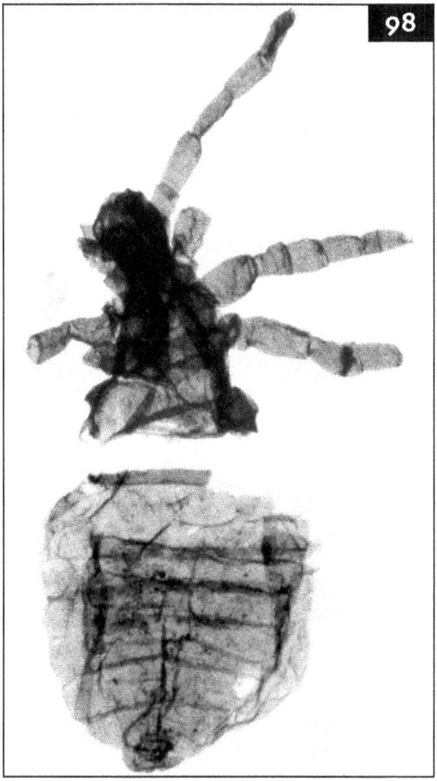

98 Tiny, juvenile trigonotarbid *Gelasinotarbus bonamoae*, nearly complete in two parts (AMNH). Body length 1.5 mm (0.06 in).

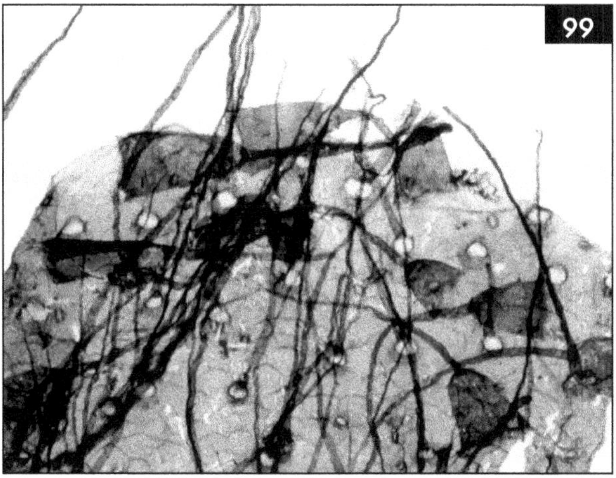

99 Piece of cuticle of the spider *Attercopus fimbriunguis* bearing characterictic silk-producing spigots. Note the bell-shaped bases of the spigots (which are derived from modified setae); ordinary setae in follicles; and tiny, elongate slit sense organs (PC). Width of slide 0.25 mm (0.01 in).

provided conclusive evidence of not only the presence of spiders but also the oldest use of silk known from the fossil record (Shear *et al.*, 1989). It was later discovered that the cuticle pattern matched other fragments already described by Shear *et al.* (1987) as a possible trigonotarbid, called *Gelasinotarbus? fimbriunguis* (**100**). Once this was recognized, many more pieces could be assembled, and much of the anatomy of the spider, now called *Attercopus fimbriunguis*, was reconstructed (Selden *et al.*, 1991). Recreating a whole animal from many tiny fragments was akin to doing a jigsaw puzzle with only half of the pieces and without the benefit of the picture on the box! Legs, which were tubular in life, were well preserved because when they were flattened, both surfaces were squashed together, strengthening the fragment; other parts such as the body, which consisted of large but single sheets of cuticle, apparently disintegrated either under pressure or during acid maceration. *Attercopus* (its name comes from *attercop* – an Old English word for a spider) is not only the oldest spider but also the only one known from the Devonian Period; other putative records of Devonian spiders were discussed and dismissed by Selden *et al.* (1991). Material of *Attercopus* has appeared in macerations from the younger South Mountain locality. Comparison with modern primitive spiders suggests *Attercopus* was a burrower and used silk to line its burrow.

Pseudoscorpions

These little animals (the largest alive today measures less than 12 mm [0.5 in] in length) resemble true scorpions in having chelate pedipalps but they lack a tail. Modern pseudoscorpions live in moss, soil, bark, under stones, and in leaf litter. They are predators on even tinier invertebrates. Until the 1980s the oldest fossil pseudoscorpion known was from Baltic amber, about 38 million years in age (Selden and Nudds, 2004, Chapter 13), so when one emerged from the acid maceration of Brown Mountain in 1990, it extended the geological range of the group 10-fold! The fossil was named *Dracochela deprehendor*, and placed in its own family, Dracochelidae, by Schawaller *et al.* (1991). *Dracochela* is so similar to living pseudoscorpions that the family seems to fit into the modern superfamily Chthonioidea (Harvey, 1992).

Mites

These minute arachnids are abundant in nearly all terrestrial environments today but rare in the fossil record. They are

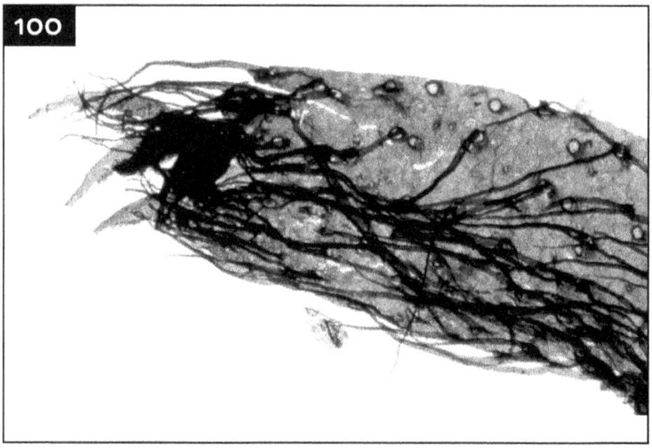

100 Tarsus (last leg segment) of the spider *Attercopus fimbriunguis*, showing characteristic fimbriate claws (upper paired claws and lower median claw are quite similar in size), setae, and slit sense organs (AMNH). Length 0.6 mm (0.02 in).

frequently overlooked because of their small size (the majority are less than 2 mm (0.08 in) long). Today's diversity of mites is second only to that of insects, with some 30,000 extant species described. The most abundant fossil mites are Pleistocene and Holocene peats and mammoth sites of the northern hemisphere, and reasonably diverse faunas in the Baltic and Dominican Republic ambers (Chapter 13). Mesozoic mites are particularly rare. The oldest known mites come from the Rhynie Chert (Selden and Nudds, 2004, Chapter 5). At Gilboa, Brown Mountain has yielded two species of oribatid mites (Norton *et al.*, 1988) and one alicorhagiid (Kethley *et al.*, 1989). A few poorly preserved specimens of oribatids similar to one of the Brown Mountain oribatid families (Devonacaridae) have been recovered from South Mountain. Oribatids are detritivores or fungivores in modern ecosystems; alicorhagiids are known to prey on nematodes. Mites play a major role in soil and litter communities today and doubtless did so in the Devonian Period.

Millipedes

Millipedes are predominantly feeders on plant detritus and live mainly in soil and leaf litter. They are the oldest land animals known from body fossils, although trace fossils attributable to land animals (including millipedes) occur in older rocks (Shear and Selden, 2001). Body fossils have been found in strata of middle Silurian age in Scotland (Wilson and Anderson, 2004). Others are known from the Devonian of Scotland and Canada (Shear *et al.*, 1996). A couple of specimens of flat-backed millipedes were reported from the Upper Devonian of the Delaware Valley (Shear and Selden, 2001), but the better known millipedes from Gilboa belong to the extinct group Arthropleurida. A new genus and species of a supposed scorpion with gills covered by circular plates (**101**), called *Tiphoscorpio hueberi*, was described by Kjellesvig-Waering (1986) from material macerated from South Mountain by Francis Hueber (1960). During a visit to Bill Shear's laboratory in Virginia to study the Gilboa arthropods, Paul Selden, then of Manchester University (UK), saw Hueber's slides of *Tiphoscorpio* fragments and noticed their similarity to a quite different animal, *Eoarthropleura*, from the Devonian of Germany (Størmer, 1976). Arthropleurids occur in the fossil record from the late Silurian to the late Carboniferous. Carboniferous arthropleurids were gigantic, possibly 2 m (7 ft) in length but, in contrast, the

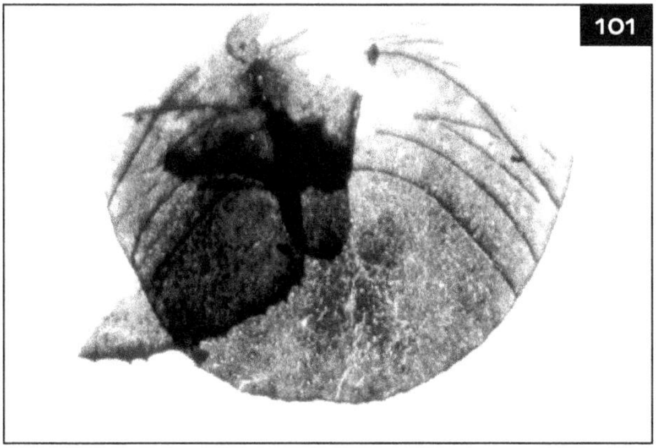

101 Plates originally described as gill covers of an aquatic scorpion but later shown to be ventral plates on the body of an arthropleurid myriapod (see text) (USNM). Width 1.6 mm (0.06 in).

Devonian ones were tiny, no more than 7–8 mm (0.3 in) long. Middle-sized eoarthropleurids, about 15 cm (6 in) in length, occur in late Silurian beds in England and Devonian rocks in Germany and North America (Størmer, 1976; Shear and Selden, 1995; Shear *et al.*, 1996). Eoarthropleurids resemble large, flat-backed millipedes, but fully articulated specimens had never been found, and nothing was known of their heads or how many segments may have been in the trunk until the tiny forms, belonging to the order Microdecemplicida, were found at Brown Mountain and South Mountain. These animals were probably less than 5 mm (0.2 in) long, and had 8 trunk segments each with two pairs of legs, a collum behind the head, a complex mouthpart system (**102**) and an anal cone (Wilson and Shear, 2000). It is likely that they were detritivores like other millipedes.

Centipedes

In contrast to millipedes, centipedes are fast-moving terrestrial predators with relatively longer legs and poison claws. Specimens belonging to the order Scutigeromorpha, to which the common house centipede *Scutigera* (**102**) belongs, have been found in rocks as old as Silurian from England (Jeram *et al.*, 1990), as well as from the Lower Devonian Rhynie Chert of Scotland (Anderson and Trewin, 2003) and the Middle Devonian Brown Mountain (Shear *et al.*, 1998). Another swift centipede from Gilboa is *Devonobius delta* (**103**), described by Shear and Bonamo

102

102 The modern house centipede *Scutigera coleoptrata*; seen here in its original habitat under a rock in the Mediterranean, it is now widespread around the world in warm climates. Body about 20 mm (0.8 in) long.

103

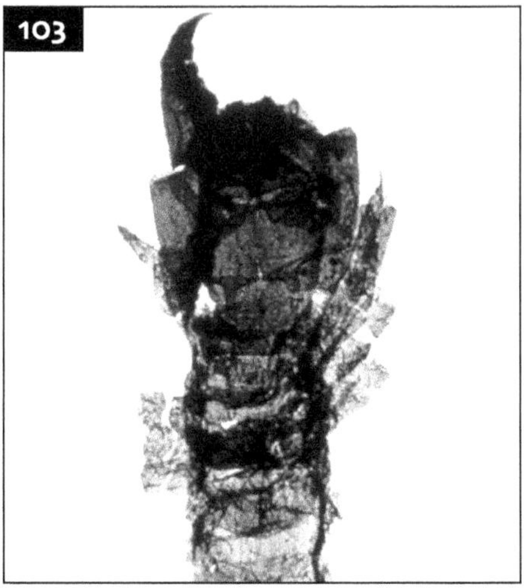

103 Front part of the centipede *Devonobius delta*, showing one of the paired poison claws on the left (AMNH). Length 1.5 mm (0.06 in).

(1988) from the Brown Mountain locality. They allocated a new order to this animal, Devonobiomorpha, which seems to be sister-group to the extant epimorphic centipedes (those in which the young hatch from the egg with the full complement of legs), and therefore more advanced than the Scutigeromorpha.

Insects and their allies

The evidence for Devonian insects is sparse. Insect precursors, such as springtails (Collembola), are known from the Rhynie Chert. Pieces of cuticle closely resembling that of modern bristletails (Archaeognatha) have been found at a number of Silurian and Devonian sites. The cuticle is thick and bears short, arched rows of oblong sockets for the insertion of scales. There is a single scrap of this distinctive cuticle from the Silurian of England, large sheets of it occur at Brown Mountain, and a few pieces from South Mountain still have the scales in place (Shear and Selden, 2001). In addition, some antenna-like fragments, possible mouthparts (**104**), and a compound eye (**105**) from Brown

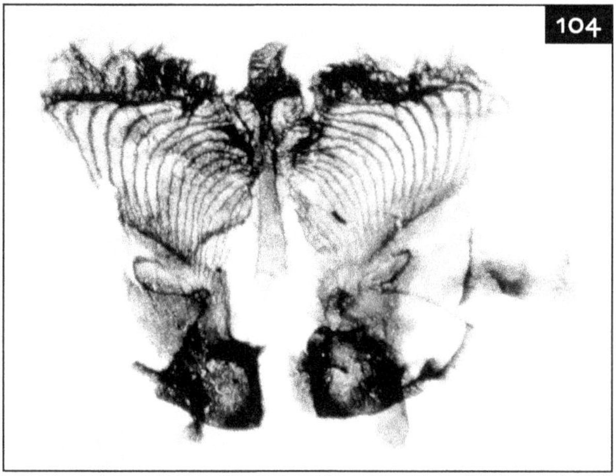

104 Enigmatic mouthparts, possibly from an insect (AMNH). Width 0.5 mm (0.02 in).

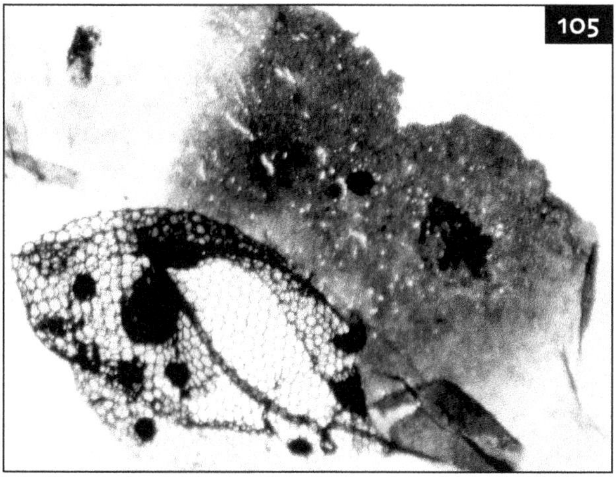

105 A scrap of cuticle bearing a modern-looking compound eye could be evidence for insects or might have belonged to one of the centipedes (AMNH). Each eye facet is 0.026 mm (0.001 in) in diameter.

Mountain may signal the presence of insects. A bristletail called *Gaspea* was macerated from Middle Devonian beds of the Gaspé Peninsula, Canada (Labandeira *et al.*, 1988). This would be the oldest insect in North America, except that Jeram *et al.* (1990) considered it was more likely to be a modern animal or molt which was in a crack in the piece of rock. The only unequivocal evidence for true insects in the Devonian is a pair of jaws called *Rhyniognatha hirsti* from the Rhynie Chert. These were first reported by Hirst and Maulik (1926); Tillyard (1928) described them and suggested that they were 'insect-like'. The specimen was studied by many experts over the years, until Engel and Grimaldi (2004) confirmed that they belong to a true insect.

Vertebrates

At South Mountain a bed, approximately 40 cm (16 in) thick, includes bones concentrated in a chaotic mass of pebbles and plant debris, including large branches up to 50 cm (20 in) long. The vertebrate material consists largely of fragmented plates of placoderms (primitive armored fish) and acanthodian (spiny shark) denticles. The discovery of a single tooth at South Mountain was exciting because, though partly pyritized, vertical striations on the lower third of the tooth represent infolding of the tooth wall – a characteristic feature of a type of tooth known as labyrinthodont. This type of tooth occurs in the earliest tetrapods – the first vertebrates on land – but also occurs in coelacanth fish. Placoderms had no true teeth, but possessed sharp-edged plates in their upper and lower jaws, which could be used for puncturing, cutting, and crushing. Acanthodians similarly lacked teeth but possessed long gill-rakers for suspension feeding. Vicky MacEwan, of the University of Manchester, UK, took the tooth to a high-resolution X-ray computed tomography unit at the University of Texas to see inside the tooth in order to determine how far into the tooth the labyrinthodont folds extended, without destroying it by sectioning (MacEwan, 2002, 2004). In tetrapods, the folds run much further into the teeth than they do in fishes. If it were tetrapod, the new record would pre-date the previous earliest known form by approximately 10 million years. The results were not conclusive, but suggested that the tooth was a little more likely to be tetrapod than fish; more material from South Mountain would be decisive in solving this tantalizing puzzle.

Paleoecology of Gilboa

The Gilboa Lagerstätte consists of a number of separate localities, each of which represent slightly different habitats (Banks *et al.*, 1985). The stump localities represent swamp forest with large trees and a considerable variety of undergrowth, judging from the number of different plants recorded: in addition to *Eospermatopteris* trunks there is the *Aneurophyton* foliage, lycopods, horsetails, and cladoxylopsids. The undergrowth most likely consisted of bushes and ground cover (creepers), all of which are extinct forms, did not have veined leaves, and reproduced by means of spores rather than seeds. So, we can envisage a forest not unlike many seen on Earth today in habitat structure but composed of very different types of plant. We have no evidence for the animal life which lived in these localities. One aspect of the plant evidence which must be borne in mind, however, is that since the forest was inundated and, eventually killed, by fast-flowing water carrying sand, some of the vegetation preserved in the sandstone might well have been transported from elsewhere by the water, and not be indicative of plants actually growing in the forest. The tree stumps are clear indicators of *in situ* preservation (i.e. they are autochthonous); the foliage debris in the sandstone might have drifted in (i.e. could be allochthonous).

At Brown Mountain, *Leclercqia* is the dominant plant: a creeping lycopod like the modern Ground Pine; *Rellimia*, a more shrubby or bushy progymnosperm also occurs there. The flora at South Mountain has a similar structure to that at Brown Mountain but consists of more, and different genera. Both seem to have been preserved in gently flowing, possibly even stagnant, water bodies within the delta environment. Of especial interest at these localities, however, are the animal remains. We find predatory arachnids (e.g. trigonotarbids, spiders, pseudoscorpions) and myriapods (*Devonobius* and scutigeromorphs), and detritus-feeders (e.g. eoarthropleurids, microdecemplicids, and mites) but no true herbivores (i.e. animals which feed on living plant material to digest).

It is most likely that the trophic system at these localities and, indeed, at all Devonian localities of early terrestrial ecosystems, differed from the more familiar ones we see today in lacking herbivory. Shear and Selden (2001) discussed this in great detail, and concluded that the lack of herbivores is most likely genuine (they have not simply all been soft-bodied and thus not preserved), but that herbivory had yet to evolve. Modern herbivores utilize a flora of fungi and other microbes in their guts to break down plant material which the animal has cut into pieces and chewed. Their guts are really fermentation chambers, and the animals rely on the flora because they lack the necessary enzymes to break down such materials as cellulose themselves. Detritus-feeders simply eat the already broken-down plant material from the forest floor, together with the decomposing microbes and fungi. We can envisage a progression from detritivory to herbivory in animals which imprison the microbes in their guts and thus bypass the external decomposition of plant material. This seems not to have occurred until much later in the Paleozoic.

Comparison of Gilboa with other early terrestrial biotas

Other famous localities preserving early terrestrial ecosystems include the late Silurian Ludford Lane, Shropshire, England (Jeram *et al.*, 1990) and the Lower Devonian Rhynie Chert of Scotland (Selden and Nudds, 2004, Chapter 5). These localities have many similarities in their trophic structure: they all lack herbivores and seem to have had a predator–detritivore food chain. There are some differences in their flora, however. The flora in the Silurian was much simpler than in the Devonian; the plants were simple, dichotomously branching axes with terminal sporangia, such as *Cooksonia*. In the early Devonian (e.g. Rhynie Chert), these simple plants were joined by primitive club-moss-like plants such as *Asteroxylon* which was a creeping form that had a main stem with side branches, all covered with scale-like 'leaves'. By the Middle Devonian, evidence from Gilboa shows that full-scale forests had developed.

Further Reading

Anderson, L. I. and Trewin, N. H. 2003. An Early Devonian arthropod fauna from the Windyfield Cherts, Aberdeenshire, Scotland. *Palaeontology* **46**, 467–509.

Arnold, C. A. 1937. Observations on fossil plants from the Devonian of eastern North America. III. *Gilboaphyton goldringiae*, gen. et sp. nov. from the Hamilton of eastern New York. *Contributions from the Museum of Paleontology, University of Michigan* **5**, 75–78.

Banks, H. P. 1966. Devonian flora of New York State. *Empire State Geogram* **4**,10–24.

Banks, H. P., Bonamo, P. M. and Grierson, J. D. 1972. *Leclercqia complexa* gen. et sp. nov., a new lycopod from the late Middle Devonian of eastern New York. *Review of Palaeobotany and Palynology* **14**, 19–40.

Banks, H. P., Grierson, J. D. and Bonamo, P. M. 1985. The flora of the Catskill clastic wedge. 125–141. In: *The Catskill Delta. Geological Society of America Special Paper* **201**. D. L. Woodrow, W. D. Sevon (eds.).

Berry, C. M. and Edwards, D. 1997. A new species of the lycopsid *Gilboaphyton* Arnold from the Devonian of Venezuela and New York State, with a revision of the closely related genus *Archaeosigillaria* Kidston. *Review of Palaeobotany and Palynology* **96**, 47–70.

Berry, C. M. and Fairon-Demaret, M. 2001. The Middle Devonian flora revisited. 120–139. In: *Plants Invade the Land: Evolutionary and Environmental Perspectives.* P. G. Gensel, D. Edwards (eds.). Columbia University Press, New York.

Dawson, J. W. 1871. On new tree ferns and other fossils from the Devonian. Quarterly *Journal of the Geological Society of London* **27**, 269–275, pl. 12.

Driese, S. G., Mora, C. I. and Elick, J. M. 1997. Morphology and taphonomy of root and stump casts of the earliest trees (Middle to Late Devonian), Pennsylvania and New York. *Palaios* **12**, 524–537.

Edwards, D., Selden, P. A., Richardson, J. B. and Axe, L. 1995. Coprolites as evidence for plant–animal interaction in Siluro–Devonian terrestrial ecosystems. *Nature* **377,** 329–331.

Engel, M. S. and Grimaldi, D. A. 2004. New light shed on the oldest insect. *Nature* **427**, 627–630.

Goldring, W. 1924. The Upper Devonian forest of seed ferns in eastern New York. *New York State Museum Bulletin* **251**, 5–92.

Goldring, W. 1927. The oldest known petrified forest. *The Scientific Monthly* **24**, 514–529.

Grierson, J. D. and Banks. H. P. 1963. Lycopods of the Devonian of New York State. *Palaeontographica Americana* **4**, 220–295.

Harvey, M. R. 1992. The phylogeny and classification of the Pseudoscorpionida (Chelicerata: Arachnida). *Invertebrate Taxonomy* **6**, 1373–1435.

Hernick, L. V. 1996. *The Gilboa Fossils.* Givetian Press, New York.

Hirst, S. and Maulik, S. 1926. On some arthropod remains from the Rhynie chert (Old Red Sandstone). *Geological Magazine* **63**, 69–71.

Hueber, F. M. 1960. *Contributions to the Fossil Flora of the Onteora 'Red Beds' (Upper Devonian) in New York State.* PhD thesis, Cornell University, Ithaca, New York.

Hueber, F. M. and Banks, H. P. 1979. *Serrulacaulis furcatus* gen. et sp. nov., a new zosterophyll from the lower Upper Devonian of New York State. *Review of Palaeobotany and Palynology* **28**, 169–189.

Hueber, F. M. and Grierson, J. D. 1961. On the occurrence of *Psilophyton princeps* in the early Upper Devonian of New York. *American Journal of Botany* **48**, 473–479.

Jeram, A. J., Selden, P. A. and Edwards, D. 1990. Land animals in the Silurian: arachnids and myriapods from Shropshire, England. *Science* **250**, 658–661.

Kethley, J. B., Norton, R. A., Bonamo, P. M. and Shear, W. A. 1989. A terrestrial alicorhagiid mite (Acari: Acariformes) from the Devonian of New York. *Micropalaeontology* **35**, 367–373.

Kjellesvig-Waering, E. N. 1986. A restudy of the fossil Scorpionida of the world. *Palaeontographica Americana* **55**, 1–287.

Kräusel, R. and Weyland, H. 1923. BeitrÑge zur Kenntnis der Devonflora. *Senckenbergiana* **5**, 154–184.

Labandeira, C. C., Beall, B. S. and Hueber, F. M. 1988. Early insect diversification: evidence from a Lower Devonian bristletail from Québec. *Science* **242**, 913–916.

MacEwan, V. K. 2002. Application of high-resolution computed tomography in palaeontology: analysis of a Middle Devonian labyrinthodont tooth from New York State, USA. *Abstracts of the 46th Annual Meeting of the Palaeontological Association, Cambridge, 2002. Palaeontological Association Newsletter* **51**, 100.

MacEwan, V. K. 2004. *Composition and Dynamics of Siluro-Devonian Terrestrial Ecosystems.* PhD thesis (unpublished), University of Manchester, UK.

Mather, W. W. 1843. Geology of New York, part I, comprising the first geological district. 301–307. In: *Natural History of New York. Part IV.* J. E. De Kay (ed.). Carroll and Cook, Albany.

Norton, R. A., Bonamo, P. M., Grierson, J. D. and Shear, W. A. 1988. Oribatid mite fossils from a terrestrial Devonian deposit bear Gilboa, New York. *Journal of Paleontology* **62**, 259–269.

Richardson, J. B., Bonamo, P. M. and McGregor, D. C. 1993. The spores of *Leclercqia* and the dispersed spore morphon *Acinosporites lindlarensis* Riegel: a case of gradualistic evolution. *Bulletin of the British Museum (Natural History)* (Geology) **49**, 121–155.

Ruedemann, R. 1926. Hunting fossil marine faunas in New York State. *Natural History* **26**, 505–514.

Schawaller, W., Shear, W. A. and Bonamo, P. M. 1991. The first Paleozoic pseudoscorpions (Arachnida, Pseudoscorpionida). *American Museum Novitates* **3009**, 1–17.

Selden, P. and Nudds, J. 2004. *Evolution of Fossil Ecosystems.* Manson, London.

Serlin, B. S. and Banks, H. P. 1978. Morphology and anatomy of *Aneurophyton*, a progymnosperm from the late Devonian of New York. *Palaeontographica Americana* **8**, 343–359.

Sevon, W. D. and Woodrow, D. L. 1985. Middle and Upper Devonian stratigraphy within the Appalachian basin. 1–7. In: *The Catskill Delta. Geological Society of America Special Paper* **201**. D. L. Woodrow, W. D. Sevon (eds.).

Shear, W. A. 2000. *Gigantocharinus szatmaryi*, a new trigonotarbid from the late Devonian of North America (Chelicerata, Arachnida, Trigonotarbida). *Journal of Paleontology* **74**, 25–31.

Shear, W. A. and Bonamo, P. M. 1988. Devonobiomorpha, a new order of centipeds from the Middle Devonian of Gilboa, New York State, USA. *American Museum Novitates* **2927**, 1–30.

Shear, W. A., Bonamo, P. M., Grierson, J. D., Rolfe, W. D. I., Smith, E. L. and Norton, R. A. 1984. Early land animals in North America. *Science* **224**, 492–494.

Shear, W. A., Gensel, P. G. and Jeram, A. J. 1996. Fossils of large terrestrial arthropods from the Lower Devonian of Canada. *Nature* **384**, 555–557.

Shear, W. A., Jeram, A. J. and Selden, P. A. 1998. Centiped legs (Arthropoda, Chilopoda. Scutigeromorpha) from the Silurian and Devonian of Britain and the Devonian of North America. *American Museum Novitates* **3231**, 1–16.

Shear, W. A., Palmer, J. M., Coddington, J. A. and Bonamo, P. M. 1989. A Devonian spinneret: early evidence of spiders and silk use. *Science* **246**, 479–481.

Shear, W. A. and Selden, P. A. 1995. *Eoarthropleura* (Arthropoda, Arthropleurida) from the Silurian of Britain and the Devonian of North America. *Neues Jahrbuch für Geologie und Paläontologie, Abhandlungen* **196**, 347–375.

Shear, W. A. and Selden, P. A. 2001. Rustling in the undergrowth: animals in early terrestrial ecosystems. 29–51. In: *Plants Invade the Land: Evolutionary and Environmental Perspectives.* P. G. Gensel, D. Edwards (eds.). Columbia University Press, New York.

Skog, J. E. and Banks, H. P. 1973. *Ibyka amphikoma,* gen. et sp. n., a new protoarticulate precursor from the late Middle Devonian of New York State. *American Journal of Botany* **60**, 366–380.

Stein, W. E., Mannolini, F., Hernick, L. A., Landing, E. and Berry, C. M. 2007. Gaint cladoxloopsid trees resolve the enigma of the Earth's earliest forest stumps at Gilboa. *Nature* **446**, 904–907.

Størmer, L. 1976. Arthropods from the Lower Devonian (Lower Emsian) of Alken an der Mosel, Germany. Part 5: Myriapoda and additional forms, with general remarks on fauna and problems regarding invasion of land by arthropods. *Senckenbergiana lethaea* **57**, 87–183.

Tillyard, R. J. 1928. Some remarks on the Devonian fossil insects from the Rhynie chert beds, Old Red Sandstone. *Transactions of the Royal Entomological Society of London* **76**, 65–71.

Wills, L. J. 1965. A supplement to Gerhard Holm's Über die Organisation des *Eurypterus fischeri* Eichw. With special reference to the organs of sight, respiration and reproduction. *Arkiv för Zoologie* **18**, 93–145.

Wilson, H. M. and Anderson, L. I. 2004. Morphology and taxonomy of Paleozoic millipedes (Diplopoda: Chilognatha: Archipolypoda) from Scotland. *Journal of Paleontology* **78**, 169–184.

Wilson, H. M. and Shear, W. A. 2000. Microdecemplicida: a new order of minute arthropleurideans (Arthropoda: Myriapoda) from the Devonian of New York State, USA. *Transactions of the Royal Society of Edinburgh: Earth Sciences* **90**, 351–375.

Mazon Creek

Chapter 7

Background: the Coal Measures

Once life became established on land it quickly developed a more complex habitat structure; as plants developed tree forms, so forests evolved (by the Late Devonian), and alongside this there was a great diversification of animal life. The late Devonian also saw the evolution of tetrapods (four-legged vertebrates) which emerged onto land in Mississipian times and started preying on the abundant invertebrate life already established there. By the Pennsylvanian there were extensive forests across the equatorial areas of the globe, which included the present areas occupied by north-west and central Europe, eastern and central USA, and elsewhere in the world such as southern China and South America. These forests are represented in the fossil record by coal seams, which are preserved because in these places the forests developed on mires, permanently waterlogged ground; the anoxic conditions prevented complete decay of the forest litter and thus prompted the formation of peat which, when compressed under a great thickness of later sediment, turned to coal. It was these vast beds of coal, and associated ironstones, clays, and other natural resources which, in Britain in particular, provided the raw materials for the Industrial Revolution.

A Coal Measure sequence of rocks presents a more complex and interesting suite of environments than a simple swamp forest; most Pennsylvanian Coal Measure sequences represent deltas with a range of environments from marine bays through brackish lagoons to sand bars, freshwater lakes, levees, and swamp forests. Delta lobes are geologically short-lived. If their sediment supply is cut off, they rapidly sink and sea water transgresses the land surface, swamping the forests. Thus, in many sequences there is a band of mud-bearing marine fossils immediately above a coal seam. The sea may persist in the area for many tens or hundreds of years before a new delta lobe builds out into the area and relatively quickly establishes new silt, then mud substrates upon which new forests can develop. Even in established swamp forests, floods are commonplace. As a result of these changing environments, Coal Measure sequences show a distinctive pattern of thin mud or shale layers, siltstones (often regularly laminated), coarse sandstones, and coal seams.

Some of the vascular plant groups discussed in Chapter 6 continued with little morphological change into the Mississippian and beyond, the bryophytes for example, while others, such as the psilophytes, gave rise to the horsetails, club-mosses, and ferns that attained gigantic proportions in the Pennsylvanian. Many of these groups formed the understorey vegetation too, together with seed-ferns, cordaites, and early conifers. Animals, too, had diversified and moved into the new niches provided by the plants. Insects had appeared, evolved wings, and some Pennsylvanian early dragonflies had a wingspan of three-quarters of a metre (30 in) (though those found in the Mazon Creek beds were smaller). Myriapods became giant in the Pennsylvanian too, with large, armoured millipedes and the enormous (>2 m [6.6 ft] long) arthropleurids, the largest known land arthropods. Vertebrates had not only followed the arthropods onto land but amphibious tetrapods had attained large size (up to 1 m [3.3 ft] long), and there were freshwater sharks with bizarre dorsal spines.

History of discovery of the Mazon Creek biota

Mazon Creek is a small tributary of the Illinois River, situated some 150 km south-west of Chicago (**106**), and it has given its name to this Lagerstätte. The fossils come principally from spoil heaps of the strip coal mines which have operated in the area over the last century. The importance of the Mazon Creek biota is that it has been so well collected, particularly by an army of keen amateurs, that it has yielded the most complete record of late Paleozoic shallow marine, freshwater, and terrestrial life. More than 200 species of plants and 300 animal species have been described, including representatives of 11 animal phyla.

Plant fossils were collected and described from natural outcrops and small mine tips in the Mazon Creek area for many years before the large Pit 11 open strip mine was opened in the 1950s. In the late years of that decade, Peabody Coal bought out the Northern Illinois Coal Company and allowed local fossil collectors to visit the pit and collect from the waste material. In strip mining, the overburden (in this case the Francis Creek Shale) is stripped off by giant buckets on drag lines to reveal the coal beneath, which is then simply dug out by smaller diggers and loaded into trucks for removal to the sorting plant. The coal is dug in long strips, so the overburden is used to back-fill the strip where coal was previously removed. It is the Francis Creek Shale which is the source of the exceptional fossils at Mazon Creek. Once news of the coal company's generosity in allowing access to its site spread, there was a regular stream of amateur fossil collectors at the site looking for the elusive special fossil.

The fossils at Mazon Creek occur in clay ironstone nodules (concretions). These usually require a winter or so of weathering before they will split easily with a single hammer-blow, usually along the weakest line which is that of the fossil. Some collectors artificially freeze and thaw the nodules to accelerate this process. Many of the concretions contain seed-fern fronds, some contain indeterminate shapes which were termed 'blobs' and consequently thrown away. Later research has shown that most of these blobs are actually fossil jellyfish, attesting to the unusual preservation of soft-bodied animals at Mazon Creek. The scramble of collectors at Pit 11 could have resulted in the loss of the best fossils to personal cabinets, never to be studied by experts were it not for the efforts of Dr ES ('Gene') Richardson who encouraged the collectors to meet at regular intervals at the Field Museum in Chicago and show their finds. In this way they learned from both the scientists and each other about what the animals and plants were and the best ways of finding them. They could be swapped, and the best specimens presented to the museum for study. The annual Mazon Creek Open House for fossil collectors continues to this day.

Some 20 years ago the Peabody Coal Company sold Pit 11 for the construction of a nuclear power plant. While mining no longer goes on there, the tips remain and are still picked over. Moreover, during construction of the power station, boreholes were drilled which passed through the Francis Creek Shale and yielded some important information about its environment of deposition.

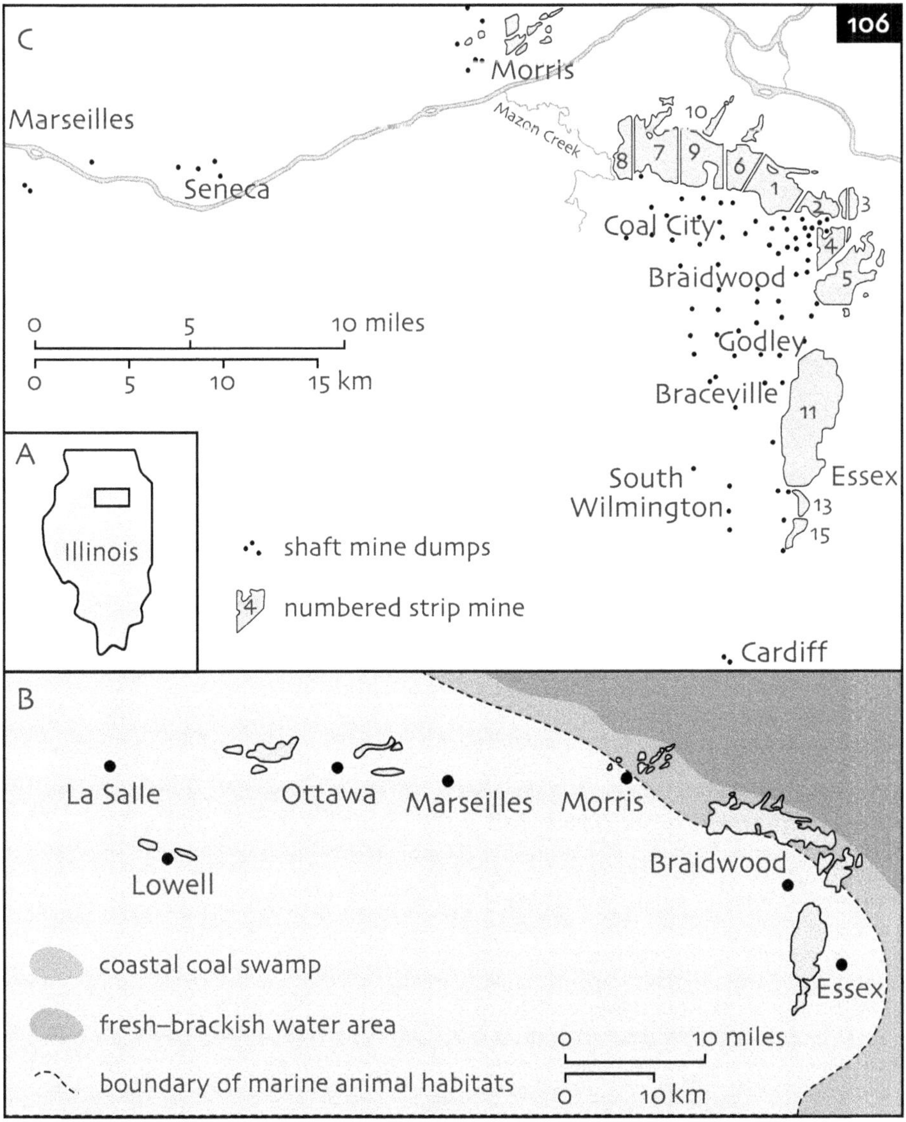

106 Locality map of the Mazon Creek area. **A**, location map, with area of marine rocks shown in blue; **B**, palaeoenvironmental reconstruction of the Mazon Creek area in relation to the strip mines; **C**, locations of the strip mines and dumps from shafts (after Baird *et al.*, 1986).

Stratigraphic setting and taphonomy of the Mazon Creek biota

The Mazon Creek fossils occur in siderite ($FeCO_3$) concretions in the Francis Creek Shale Member of the Carbondale Formation, of Westphalian D age. The Francis Creek Shale overlies the Colchester N° 2 Coal Member, and is itself overlain by the Mecca Quarry Shale Member (**107, 108**). The Colchester Coal is generally about 1 m (3.3 ft) thick; the Francis Creek Shale is typically a grey, muddy siltstone with minor sandstones and varies from complete absence up to 25 m (82 ft) or more in thickness. The siderite concretions occur only in the lower 3–5 m (10–16 ft) of the member, and only where the shale is more than 15 m (50 ft) thick. The shale is coarser and bears sandstones near its top. The Mecca Quarry Shale is a typical Pennsylvanian black shale that peels easily into sheets, and contains a rich fauna of sharks and their coprolites, exhaustively monographed by Zangerl and Richardson (1963). It is generally about 0.5 m (18 in) thick but is absent over areas where the Francis Creek Shale is more than about 10 m (33 ft) thick, and thus also over the nodule-bearing parts of the Francis Creek Shale.

The fossils are found almost only within the siderite concretions. When these are broken open they reveal a nearly three-dimensionally preserved organism, though the fossil becomes more flattened towards the edge of the nodule. Fossils are generally preserved as external molds, commonly with a carbonaceous film (if a plant). There may be crystals such as pyrite, calcite, or sphalerite on the mold surfaces. The commonest mineral on these surfaces, however, is kaolinite, a white, soapy, clay mineral. This is often found completely infilling the space between the molds (i.e. forms a cast). It is soft and therefore easy to remove mechanically. Fossils with few hard parts or little rigidity, such as jellyfish, may collapse completely and be preserved as composite molds; even arthropods may show dorsal and ventral structures superimposed.

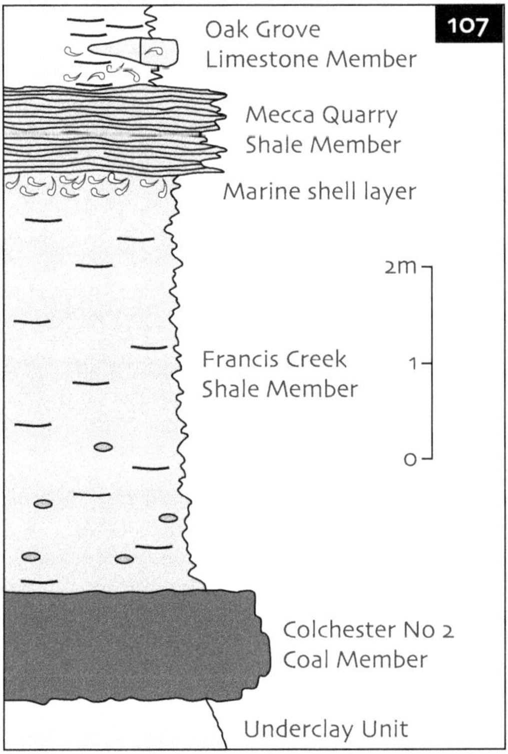

107 Stratigraphic log of the Francis Creek Shale and associated Members of the Carbondale Formation, Illinois (after Baird *et al.*, 1986).

108 Disused Pit 8 in the Colchester N° 2 Coal. The spoil forming the raised banks is the source of the Mazon Creek nodules.

Most fossils show very little decay and, indeed, there are instances of bivalves preserved on the edge of the nodule at the end of their death trail (**109**)! The concretions do not normally extend far beyond the fossil, so the size and shape of the nodule correlates with the organism inside. Few concretions exceed 300 mm (12 in), so large animals are rare in the biota. The evidence presented so far tends to indicate that the concretions formed very soon after death and burial of the organisms, mollusks stopped in their tracks, seed-fern pinnules at right-angles to the bedding, and little decay (**110**). Large animals such as big fish and amphibians, it is presumed, could escape this environment and not be preserved. That the organisms are preserved three-dimensionally, at least in the cores of the nodules, while the surrounding matrix, like all siltstones, is greatly compressed implies that the concretions formed before any appreciable compaction. Indeed, the siltstone laminae can be seen (**111**) to widen progressively from the matrix towards the centre of the nodules, suggesting that the concretions grew during compaction. Also, cracks within the nodules, commonly infilled with kaolinite, can be related to dewatering of the sediment (syneresis) during their formation.

Because the fossil-bearing concretions mirror the shape of the fossil, and the fossils are generally situated fairly centrally within the concretions, we can assume that the organisms contributed significantly to their formation. Moreover, barren nodules can usually be explained as containing rather flimsy fossils, unrecognizable organic matter, trace fossils, and the like. The nodules contain about 80% siderite cement, which implies that when the concretions formed there was at least 80% water by volume in the sediment before compaction. Iron would normally react with sulfur under the influence of anaerobic bacteria in the presence of decaying organic matter to form pyrite (FeS_2), as in the Hunsrück Slate (Selden and Nudds, 2004, Chapter 4), in preference to siderite but once any sulfate was used up by this process (some pyrite does occur in the nodules), then methanogenic bacteria would help to generate siderite. Conditions at Mazon Creek which helped this process would have been an abundance of iron and a weak supply of sulfate. The Mazon Creek nodules are commonly asymmetrical, with flatter bottoms and more pointed tops. This is due to the effects of gravity: the weight of the carcass presses into the sediment beneath, the concretion can grow more easily upwards (where there is less compaction), and any light fluids resulting from decay would also rise preferentially.

Description of the Mazon Creek biota

The Mazon Creek biota actually consists of two biotas: the Braidwood and the Essex; the former occurs mainly in the north of the area, the latter in the south. Eighty-three percent of the Braidwood nodules contain plants, with the next commonest (7.8%) inclusions being coprolites. Following these (in descending order) can be found: freshwater bivalves (1.8%), freshwater shrimps (0.5%), other mollusks (0.4%), horseshoe crabs (0.3%), millipedes (0.1%), and fish scales (0.1%); insects, arachnids, fish, and centipedes form the remainder (<0.1%). On the other hand, only 29% of the Essex biota consists of plants, the commonest animal being the 'blob' *Essexella* (42%). Following these (in descending order) are burrows and trails (5.9%), the marine solemyid bivalves (5.5%), coprolites (4.8%), worms (2.8%), miscellaneous mollusks (1.9%), the marine shrimp *Belotelson* (1.8%), the marine bivalve *Myalinella* (1.4%), miscellaneous shrimps (0.5%), the crustacean *Cyclus* (0.5%), the enigmatic Tully Monster (0.4%), the scallop *Pecten* (0.3%), the jellyfish *Octomedusa* (0.3%), and miscellaneous fish (0.2%); insects, millipedes and centipedes, hydroids, horseshoe crabs, arachnids, and amphibians form the

109 Death trail of undescribed solemyid bivalve. The bivalve was still alive and attempting to escape while the siderite nodule was forming, resulting in a fossilized trail and the bivalve at the edge of the nodule. (CFM). Scale bar is in cm.

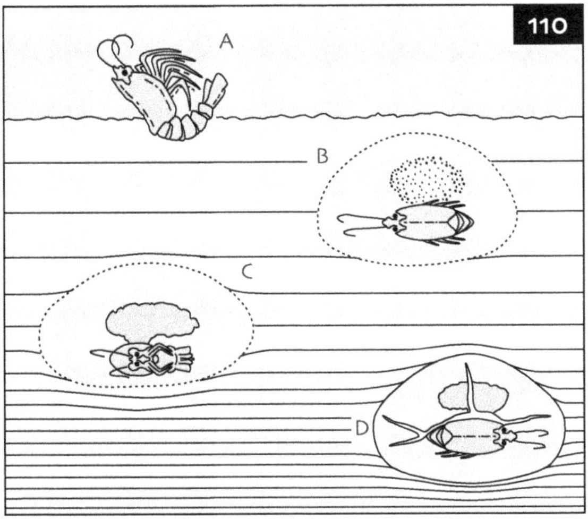

110 Diagram showing the rapid formation of a siderite nodule around a dead shrimp. **A**, Dead shrimp lands on sea floor; **B**, partial decay by bacteria, volatiles rise; **C**, siderite precipitation while compaction begins; **D**, further compaction of surrounding sediment while dewatering (syneresis) causes cracks to extend from centre of nodule outwards (after Baird *et al.*, 1986).

111 Section through the Francis Creek Shale laminites and part of a siderite nodule. Note the laminae widen towards the nodule, indicating greater compaction of the surrounding silt/clay laminites than the nodule, and that compaction had started during growth of the nodule. Thickest part of specimen (left) is 1.6 in (4 cm).

remainder (< 0.1%). From this list it can be seen that the Braidwood biota consists of terrestrial and freshwater organisms while the Essex biota consists of predominantly marine organisms, with some drifted plants etc. No marine organisms can drift into freshwater but freshwater and terrestrial organisms can be washed down into the sea. Notice, too, that the marine Essex biota is not a typical marine biota (there are no brachiopods, corals, crinoids), so it must represent a reduced-salinity, and perhaps muddy, environment which fully marine animals cannot tolerate.

Plants

The Mazon Creek nodules preserve a typical Coal Measure flora, exceptional only in that its preservation is better in the concretions than in other Pennsylvanian clayrocks. Large fossils are not preserved in the Francis Creek Shale but the presence of tree-sized club-mosses and horsetails can be inferred from pieces of bark (*Lepidodendron* and *Calamites*, respectively) and foliage (*Lepidophylloides* and *Annularia*, respectively). Seed-fern pinnules are the commonest fossils in all Mazon Creek nodules; examples include *Neuropteris* (**112**), *Pecopteris* (**113**), and *Alethopteris* (**114**). Comparison with modern coastal swamp forests suggests that much of the plant debris found in the Mazon Creek nodules originated not from the forest itself but was carried down by streams from far inland, so possibly represents a mixture of coastal and upland forest. Plant debris is allochthonous (i.e. drifted from its original life position) in both Braidwood and Essex biotas, but is commoner in the former.

Cnidarians

This phylum includes the corals and anemones (Anthozoa), hydroids (Hydrozoa), Scyphozoa and Cubozoa (jellyfish), and a few other groups. Apart from the corals, they have no mineralized skeletons but some jellyfish have stiffened parts. Most 'blobs' in Mazon Creek nodules are jellyfish remains, and some show distinct tentacles and other structures. *Essexella*, for example (**115**), shows a bell with a cylindrical sheet hanging below it. *Essexella* belongs in the Scyphozoa, while *Anthracomedusa* (**116**), with four bunches of numerous tentacles, is a cubozoan.

112 The seed-fern *Neuropteris*. (MU). Pinnule is 2.4 in (6 cm) long.

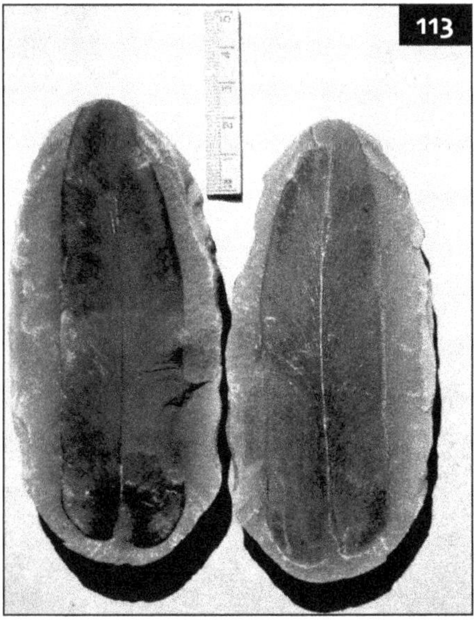

113 The seed-fern *Pecopteris*. (MU). Scale bar is in cm.

114 The seed-fern *Alethopteris*. (MU). Nodule is 2.7 in (7 cm) long.

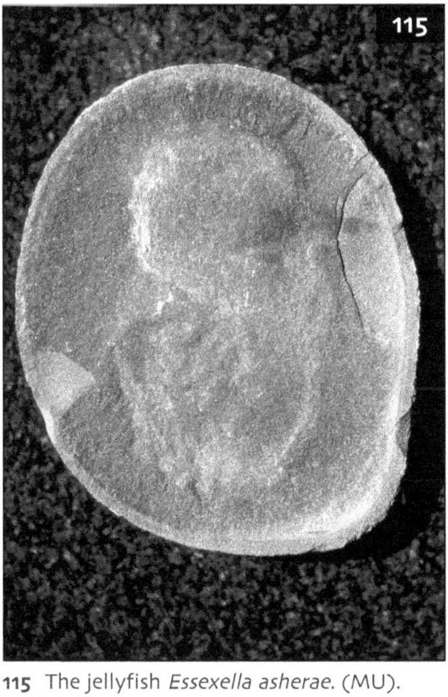

115 The jellyfish *Essexella asherae*. (MU). Nodule is 2.4 in (6 cm) long.

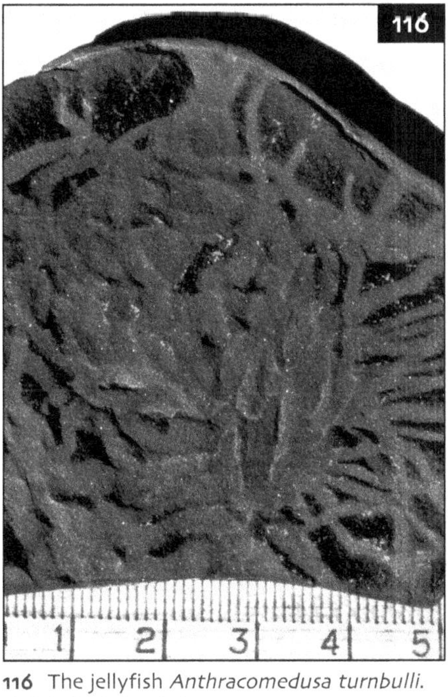

116 The jellyfish *Anthracomedusa turnbulli*. (MU). Scale bar is in cm.

Bivalves

Bivalves are common fossils on account of their hard, calcareous shells (valves), but those in the Mazon Creek nodules are important because they commonly preserve soft-part morphology. There is a high diversity of bivalves in the Mazon Creek, with 12 superfamilies represented. It is convenient to divide the fauna into freshwater and marine forms: i.e. those found in the Braidwood and Essex biotas respectively. The commonest marine bivalve occurs as articulated valves (called 'clam-clam' specimens by amateur collectors), and is occasionally found at the end of an unsuccessful escape trail (**109**). Previously misidentified as *Edmondia*, it has been shown recently to be an undescribed solemyid; true *Edmondia* are rare at Mazon Creek. These solemyids are burrowers (infaunal benthos), but other common bivalves in the Essex biota are the thin-shelled swimmers (nekton) *Myalinella* and *Aviculopecten* (**117**). The family Myalinidae also includes freshwater forms such as *Anthraconaia*, found in the Braidwood biota.

Other mollusks

Three other classes of mollusk occur in the Mazon Creek biota: Polyplacophora, Gastropoda, and Cephalopoda. Though there are many freshwater gastropods (snails) today, the only gastropods at Mazon Creek are from the marine Essex biota. Polyplacophora (chitons) are exclusively marine animals which are rare in the fossil record because they usually inhabit rocky shores. However, one genus, *Glaphurochiton* (**118**), occurs in the Essex biota. Cephalopods are also wholly marine and, though normally common in fully marine Pennsylvanian rocks, they are rarer than chitons at Mazon Creek, but quite diverse in the Essex biota. In addition to the orthocone bactritoids, coiled ammonoids and nautiloids with external shells, coleoids (with internal hard parts) are also present. *Jeletzkya* is a small, squid-like coleoid with an internal shell similar to a cuttlebone.

Worms

Apart from their tiny jaws, called scolecodonts, polychaete annelids are rarely preserved as fossils because they are soft-bodied. Therefore, the great diversity of polychaetes found at Mazon Creek is an important contribution to the fossil record of this important group of marine animals. One of the commonest Mazon Creek polychaetes is *Astreptoscolex* (**119**), which shows a segmented body and short chaetae (spines) along each side.

117 The bivalve *Aviculopecten mazonensis.* (MU). Scale bar is in cm.

118 The chiton *Glaphurochiton concinnus.* (MU). Scale bar is in cm.

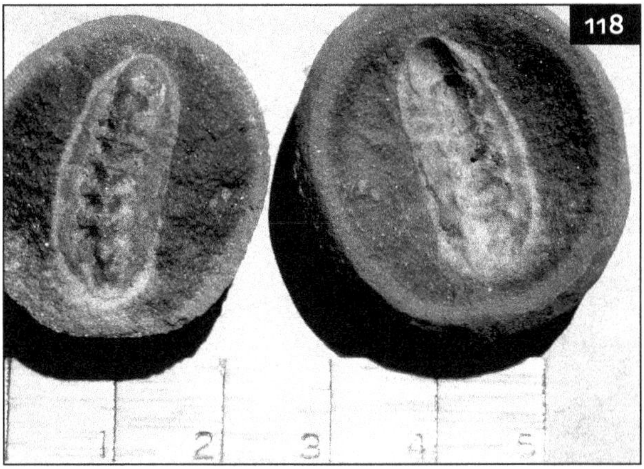

119 The polychaete annelid *Astreptoscolex anasillosus.* (MU). Scale bar is in cm.

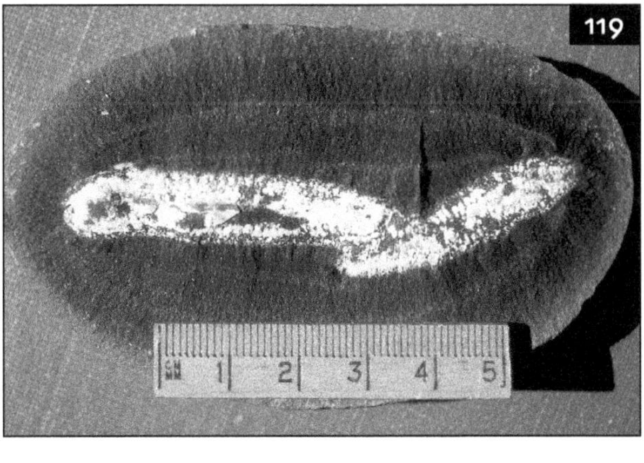

Shrimps

A wide variety of Crustacea occur at Mazon Creek, many of which have a shrimp-like body shape. There are both freshwater and marine shrimps, found in the Braidwood and Essex biotas, respectively. *Belotelson magister*, a robust species, is by far the most common shrimp in the marine Essex biota. The second commonest marine shrimp is *Kallidecthes*. *Acanthotelson* (**120**) and *Palaeocaris* are freshwater shrimps which occur in the Braidwood biota and are also found rarely in the Essex nodules, where they presumably were washed down by currents.

Other crustaceans

Some squat, crayfish-like forms also occur at Mazon Creek: *Anthracaris* in the Braidwood biota and *Mamayocaris* in the Essex biota. There are also phyllocarids (*Dithyrocaris*) in the Essex biota, conchostracans (crustaceans almost completely enclosed in a bivalved carapace) which were probably fresh- or brackish-water forms, some ostracodes and barnacles. A crustacean which commonly occurs in Pennsylvanian nodules, *Cyclus*, is also found in the Essex biota. As its name suggests, *Cyclus* has a round, dish-like carapace, and has been thought of as a fish parasite, though it may have been free-living.

Chelicerates

The arthropod subphylum Chelicerata includes horseshoe crabs (Xiphosura), scorpions, eurypterids, spiders, mites, and other arachnids. Mazon Creek has yielded some exceptionally fine fossils of these animals, which have provided a great deal of information on the evolutionary history of the chelicerates. *Euproops danae* (**121**) is one of the best known horseshoe crabs in the fossil record. Work by Dan Fisher (1979) of the University of Michigan revealed the amphibious mode of life of these animals, which are found predominantly in the Braidwood biota.

Eurypterids were discussed in Chapter 5 in connection with the Bertie Waterlime. By the Pennsylvanian they were mostly amphibious and represented largely by the genus *Adelophthalmus*, with numerous specimens in Mazon Creek nodules.

Terrestrial arachnids are well represented in the Braidwood biota, and the extinct group Phalangiotarbida has the most numerous representatives; trigonotarbids are also well represented. The latter are close to spiders in morphology but lack poison glands and silk and, as arachnids go, are relatively common in late Paleozoic terrestrial ecosystems. Two living orders of arachnids, Uropygi (whip scorpions) and Amblypygi (whip spiders), have some well-preserved examples in the Braidwood biota (**122**). An interesting order of living arachnids, Ricinulei, is represented at Mazon Creek by three genera. Ricinulei are rarely encountered animals, even today, and are restricted to tropical forests and caves. They are only known from the Pennsylvanian and Recent, but seem to have changed little in between. Three other arachnid orders occur at Mazon Creek: Opilionida (harvestmen), Solpugida (camel spiders) and Scorpionida. The scorpions are the most ancient of arachnids and, though exclusively terrestrial today, are found in aquatic environments in the Silurian. Scorpions are found only in terrestrial environments by the Pennsylvanian, and are the second most abundant arachnid in the Braidwood biota.

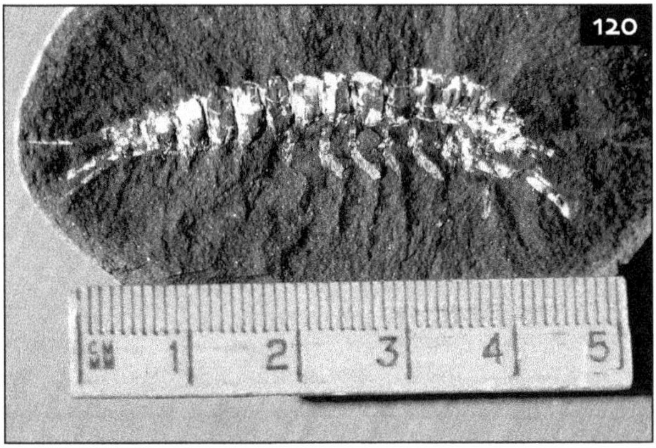

120 The shrimp *Acanthotelson stimpsoni*. (MU). Scale bar is in cm.

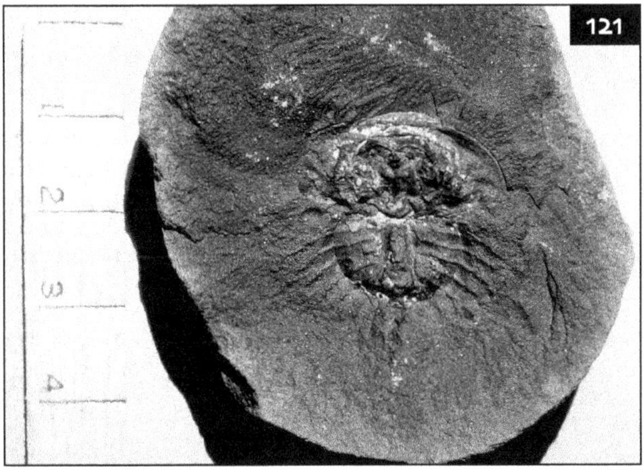

121 The horseshoe crab *Euproops danae*. (MU). Scale bar is in cm.

122 Cast of uropygid arachnid *Geralinura carbonaria*. (CFM). Scale bar is in mm.

Insects

Six orders of insects occur at Mazon Creek, all of which are now extinct. Much of our knowledge of Pennsylvanian insects comes from the 150 species found in Mazon Creek nodules. Palaeodictyoptera were medium to giant flying forms with patterned wings and comprised half of all known Paleozoic insects. Both nymphs and adults were terrestrial and had sucking mouthparts. Megasecoptera were similar to Palaeodictyoptera but had more slender, often petiolated (stalked) wings. Diaphanopterodea resembled Megasecoptera with the one exception that they could fold their wings over their backs like modern butterflies and damselflies do. Protodonata, as their name suggests, were distantly related to modern Odonata (dragonflies and damselflies). Some reached giant proportions in the Pennsylvanian and Permian, with wingspans of 710 mm (28 in). It is presumed that, like Odonata, the nymphs led an aquatic life, but none have been found.

'Protorthoptera' is a general name given to a large group of Paleozoic insects which resemble the modern Orthoptera (grasshoppers, katydids, locusts, and crickets) but lacked jumping legs. They form the stem-group which gave rise to modern Orthoptera, Dermaptera (earwigs), Phasmatodea (mantids), and some other extinct and living orders. Twelve families occur at Mazon Creek, and *Gerarus* is the commonest insect fossil found there. *Gerarus* belongs to the group which is thought to have given rise to the giant Triassic Titanoptera, now extinct. It was once thought that the familiar roaches occurred in rocks as old as Pennsylvanian because of the abundant remains of roach-like animals in Pennsylvanian nodules. However, it has now been shown that these animals possess some rather primitive traits, such as a large external ovipositor which evolved even before the origin of flight, so these stem-group Dictyoptera are better described as 'roachoids' or Blattodea (Grimaldi and Engel, 2005). In the nodules from Mazon Creek (**123**) their veined wings are often mistaken for seed-fern pinnules.

Myriapods

Myriapods are multilegged arthropods which include the centipedes (Chilopoda), millipedes (Diplopoda), two other living classes (Symphyla and Pauropoda), and the Paleozoic Arthropleurida. Myriapods were among the earliest known land animals (Chapter 6), but by Pennsylvanian times some had become gigantic, and many sported fierce spines, presumably for defence against predators. Short millipedes which could roll up into a ball (Oniscomorpha: *Amynilyspes*) were present in the Braidwood biota, but the most dramatic forms belong to the extinct order Euphoberiida. *Myriacantherpestes* (**124**) probably reached around 30 cm (1 ft) in length and had long, forked, lateral spines and shorter dorsal spines. *Xyloiulus* (**125**) was a more typical, cylindrical, spirobolid millipede. Millipedes are generally detritus feeders, while centipedes are carnivorous.

Centipedes at Mazon Creek include the scolopendromorph *Mazoscolopendra* and the fast-running scutigeromorph *Latzelia*. Arthropleurids range from tiny forms in the Silurian to the Pennsylvanian, when they became the largest known terrestrial animals, reaching 2 m (6.6 ft) in length. Nevertheless, like millipedes, they were probably detritus-feeders. Isolated legs and plates of *Arthropleura* occur at Mazon Creek. Onychophora (velvet worms) should also be mentioned here. They are known from the Cambrian (e.g. *Aysheaia*, Chapter 3), when they were marine, to the Recent, when they are wholly terrestrial. The Mazon Creek *Ilyodes* was collected from natural outcrops and it is not known whether it came from the terrestrial Braidwood or marine Essex biota.

For Product Safety Concerns and Information please contact our EU
representative GPSR@taylorandfrancis.com
Taylor & Francis Verlag GmbH, Kaufingerstraße 24, 80331 München, Germany

www.ingramcontent.com/pod-product-compliance
Ingram Content Group UK Ltd.
Pitfield, Milton Keynes, MK11 3LW, UK
UKHW022154190926
475702UK00003B/27

* 9 7 8 1 1 3 8 4 2 4 0 8 1 *